DICTIONARY
for MANAGING TREES
in URBAN ENVIRONMENTS

Danny B Draper and Peter A Richards

CSIRO
PUBLISH'

INSTITUTE OF AUSTRALIAN

CONSULTING ARBORICULTURISTS

National Library of Australia Cataloguing-in-Publication entry
    Dictionary for managing trees in urban environments/editors: Danny B Draper, Peter A Richards.

    9780643096073 (pbk.)

    Includes bibliographical references and index.

    Urban forestry – Dictionaries.
    Trees in cities – Dictionaries.

    Draper, Danny B.
    Richards, Peter A.

    635.97703

Published by
CSIRO PUBLISHING
150 Oxford Street (PO Box 1139)
Collingwood VIC 3066
Australia

Telephone: +61 3 9662 7666
Local call: 1300 788 000 (Australia only)
Fax: +61 3 9662 7555
Email: publishing.sales@csiro.au
Web site: www.publish.csiro.au

Front cover image by iStockphoto

Set in 11/15.5 Adobe Times New Roman PS and Myriad MM
Cover and text design by James Kelly
Typeset by Desktop Concepts Pty Ltd, Melbourne
Printed in Australia by Ligare

PEFC™
PEFC/21-31-17

The book has been printed on paper certified by the Programme for the Endorsement of Forest Chain of Custody (PEFC). PEFC is committed to sustainable forest management through third party forest certification of responsibly managed forests.

CSIRO PUBLISHING publishes and distributes scientific, technical and health science books, magazines and journals from Australia to a worldwide audience and conducts these activities autonomously from the research activities of the Commonwealth Scientific and Industrial Research Organisation (CSIRO).

The views expressed in this publication are those of the author(s) and do not necessarily represent those of, and should not be attributed to, the publisher or CSIRO.

FOREWORD

We live in a time of increasing urbanisation. Indeed, world urbanisation is predicted to exceed 66% by 2025. The movement of people from rural areas to cities during the 20th century has been called the largest human migration in history. As such, the urban environment is of increasing importance to more and more people. There are, of course, trees and other vegetation alongside people in many urban areas but only recently, in about the last 20 years, have these other living components in urban areas received serious attention. For a long time, 'cities' and 'nature' were seen as mutually exclusive, a view which was reinforced by the removal or covering over of many natural sites and the predominance of built infrastructure in cities. In making cities, however, people often retain some pre-existing trees and vegetation and also introduce new trees and vegetation, often for ornamental reasons and to enhance recreational spaces. It is the tree component of these urban environments that is the focus of this dictionary.

Urban trees, both individually and collectively (together with other woody plants) as the urban forest, are the most conspicuous elements of 'nature' in urban areas. They are symbols of green, healthy cities and have the potential to play a key role in providing high quality urban environments. The benefits of trees in urban environments are many and varied. They include environmental and ecological benefits, physiological benefits for humans, social and psycho-logical benefits for individuals and communities, aesthetic benefits and eco-nomic benefits for businesses, people and communities. It is only in about the last 20 years that research into these benefits has begun and many people are not yet fully aware of the breadth of the benefits provided by urban trees. Nevertheless, enough people are aware of the importance of urban trees to have led to a greater recognition of them, as well as increased discussion about and reporting on issues concerned with urban trees.

Nowadays, urban ecosystems, in which urban trees play a key role, are also receiving attention in their own right. Indeed, many people believe that our 21st century cities need to function in more ecologically sustainable ways and that the ecological functioning of cities is as important as other aspects of cities. Urban ecosystems are recognised as being created by humans specifically for dwelling and any serious study in the field of urban ecology needs to be multidisciplinary, bringing together the natural, physical and social sciences. An understanding of urban tree biology is critical to understanding urban ecosystems.

It is in this context that tree management in urban environments occurs. Trees in urban areas are receiving greater attention than ever before, yet our knowledge and understanding of them has only developed relatively recently. There is an urgent need to bring together the necessary information to provide a basis for meaningful communication about urban trees. This communication involves many people, from tree experts, to other professionals and members of the wider community.

Urban tree managers are charged with the responsibility of managing trees to provide the maximum benefits, across a wide range of areas, for the minimum costs. Urban tree management is complex and involves the biology of the trees themselves (as living plants), the physical environments in which they are found (including soils and nearby built infrastructure), interactions with other organisms (such as fungi, insects and vertebrates), aspects of city planning and legal systems and people's perceptions of trees in their living and working environments.

This dictionary brings together, in one concise document, terms used by those dealing with the management of trees in urban environments. It recognises that many different people, with varying educational backgrounds, have a need to understand technical information about trees when involved in decision-making about urban trees. The dictionary aims to provide a comprehensive, stand-alone reference to assist people to understand structural and functional aspects of urban trees and how these need to be considered when decisions about trees are made.

The dictionary should become an essential reference for those professionals whose work involves urban trees, either at the core of their professional practice or as one of the aspects encountered in or impinging upon their area of professional expertise. The definitions, descriptions and diagrams of terms in this dictionary will assist professional tree consultants and managers in the preparation of accurate written reports and other documents about urban trees. It will also help other professionals, without a background in tree biology, to understand such documents and develop their knowledge of urban trees and the principles underlying their management. A subset of the terms, appended to tree reports or management documents, should also assist members of the community to understand these documents so that they are better able to contribute to the appropriate management of our urban trees.

Dr Jane Tarran
Faculty of Science
University of Technology Sydney, Australia

CONTENTS

LIST OF CONTRIBUTORS

Peter Castor, Lawrence Dorfer, Danny Draper, Jonathan Epps, Dr Peter Nichols, Peter Richards and Neville Shields – members of IACA Technical Committee No. 2.

Anthea Stead Smith – graphic design of the IACA logo.

Craig Parkinson – illustrations as pastel or greyscale.

ACKNOWLEDGEMENTS

To Donna Draper, Angela Draper, Sarah Draper, Linda Richards, Adelaide Richards and Millie Richards for their support and patience over a long period during the preparation of this work.

Christopher Byron for advice on the topic Excavation and Construction; Bronwen Castor; Mark Felgate for support of the project since its inception and thorough review of the draft; David Ford; Dr Paul Ganderton; Jonathan Hobbs; Paul Laverty; Dennis Marsden; David Reiner for advice on the topic Excavation and Construction; Duncan Slater of Myerscough College at University of Central Lancashire; Sue Slaytor; Dr Jane Tarran for undertaking the academic review of this work, for guidance and for preparing the foreword; David Thomas for advice on the topic Planning and Law; Warwick Varley; Perry Ward for advice on the topic Soils; and Sue Wylie.

INTRODUCTION

It is the intent of this dictionary to provide a terminology universal to Arboriculture and Urban Forestry. The dictionary contains as complete a list as possible of words and terms used in the management of urban trees to further the understanding of Arboriculture and Urban Forestry and the development of the Arboricultural profession. This will benefit public interest and those within the community charged with making decisions about urban trees. It will facilitate better communication about Arboriculture and Urban Forestry. Many of the terms are from Arboricultural science, while others are empirical in origin; some are derived from unproven but commonly applied concepts. It is the role of this dictionary to record these terms and their meanings but this generally does not extend to analysis or critique of their usefulness.

A growing awareness of the importance of urban trees is being realised through research into Arboriculture, Urban Forestry and related disciplines. This research appears in both scientific journals and texts. The texts are usually specialised and contain many new concepts, terms and phrases with most including a glossary defining their key words. As this body of works grows, it reveals the broad range of related terms required to understand the subject area. There are, however, areas of uncertainty in the terminology as used in many of the texts. Sometimes concepts are only partly defined and so uncertainty can remain as to their exact meaning. Here the dictionary aims to provide clarity. Some new concepts have been introduced where the existing terminology to describe trees was limited or nonexistent.

This dictionary has been compiled to allow for broad application and use by a wide variety of people. It therefore endeavours to convey, in plain language, concepts which are sometimes complex. This has been achieved by cross-references between most major terms. To assist further, some diagrams have been provided. Botanical terms pertinent to an understanding of Arboriculture and Urban Forestry have been included while omitting the many terms readily found within dictionaries of general plant sciences and botany.

The value of language to describe accurately a tree or a tree problem can never be underestimated. Such accuracy is critical to successful management of trees in urban environments. A photograph of a tree or tree problem can be an invaluable resource but an inability to interpret and describe the image will diminish its worth considerably. A correct description of the image is essential for the image to be useful.

The following list of words and phrases is by no means exhaustive, but every effort has been made to make this dictionary comprehensive – its development is expected to be a continuing process. It is envisaged that this dictionary will assist in the preparation of reports for the management of trees, procedures and planning instruments such as Tree Management Policies and Tree Management Orders and promote a greater understanding of Arboriculture and Urban Forestry.

The reference to any proprietary products in this dictionary is under no circumstances to be interpreted as an endorsement of that product or business.

HOW THIS DICTIONARY WORKS

For each entry the subject word/phrase is listed in bold followed by the definition, for example:

Branch stub A section of branch remaining beyond the branch collar and usually evident after a *lopping* episode or as a result of *branch failure*, or *natural pruning.*

Where possible, each definition provides a primary definition – a simple and concise meaning. However, in some entries the primary definition is followed by a more detailed description where this is considered appropriate to explain complex concepts. Where a word/phrase has more than one meaning, those meanings are listed.

Where a word/phrase within a definition appears in italics, this indicates that a separate definition for that term is listed within the dictionary. This will assist readers with cross-referencing and they can also consult the index of topics. Where a word/phrase is defined elsewhere in the dictionary, but not italicised within a definition, its connection is not considered significant or is peripheral to the word/phrase being defined.

Where a word/phrase is considered synonymous with another term in the dictionary, 'See' plus the primary term is given, for example:

Lop See *Lopping.*

The main term or phrase is the one most commonly used or which most accurately represents the concept. Where a closely related word/phrase or its antonym is listed, the term/phrase will be followed by 'See also' plus the appropriate term/phrase, for example:

Acoustic resonance Auditory reverberation within an object and the air after an object has been struck. See also *Sounding.*

Where a word/phrase is used often and is known by its acronym, this has been included wherever possible, for example:

Leaf area index (LAI), First order structural branch (FOSB).

Diagrams, sketches, tables and photographs are provided to further demonstrate and complement the meaning of the word/phrase located nearby. Many diagrams are labelled simply and must be considered in conjunction with the definition of the association word/phrase. This will give the reader a better understanding of the concept.

All references are cited to enable and encourage further research by the reader.

To help the reader locate all the words or phrases commonly associated with a particular subject area, an index of topics has been created. The index has been further enhanced in that, where sets of words are connected to a quantitative or qualitative description, they have been grouped in ascending order following the primary definition rather than being in alphabetical order, for example:

Deadwood, Small deadwood and *Large deadwood.*

Seeds

Ignorance
Propagates mistrust and suspicion,
Prejudice denies all hope of reason
Fear and bigotry are colluded,
Imprisoning a diminishing reality
Depauperate and deluded.

Knowledge
Supplants fear of the unknown,
Illuminates for a courageous mind
To grow and not be frightened,
As it journeys an expanding world
Empowered and enlightened.

Wisdom
Knows power an unstable device,
Balancing successes allure with humility
Admits sum to potential a pitiful load,
And strength a force to elevate fellows
To engage an endless road.

DBD

Abatement law Law/s developed to avert or remedy *nuisance*.

Aberrant Not representative of normal *form*, *atypical*, unusual and not indicative.

Abiotic Not living, e.g. wind, rain, fire, light, snow, temperature and moisture extremes. See also *Biotic*.

Abiotic factors Non-living influences. 1. Environmental factors, e.g. wind, rain, fire, light, snow, temperature and moisture extremes. 2. Mechanical factors, e.g. root severance. 3. Chemical factors, e.g. high soil salinity and *phytotoxic* pesticides (Costello *et al.* 2003).

Abnormal vigour See *High vigour*.

Abrasion wound *Mechanical wound* causing *laceration* of tissue by an abrasive impact *episode,* e.g. grazed by a motor vehicle or the continuous action of the rubbing of *crossed branches* or stems where no graft has formed.

Abrupt changes Sudden changes rather than those occurring gradually, e.g. broken water pipes saturating soil, gas leaks.

Abscise To throw off or to shed.

Abscission Shedding of plant organs, e.g. fruit, leaves or branches, usually when the organ is mature or aged, by the formation of a corky layer across its base. This may be influenced by stress, drought (to prevent moisture loss), senescence, declining condition, reduced vigour and also occurs annually in deciduous trees.

Absorbing roots See *Fine roots*.

Acaulescent A *trunkless* tree or a tree supported by a very short *trunk*. See also *Caulescent*.

Accelerated growth The concept where large *xylem* formed at *wound margins* of wounded *sapwood* develops at a rate faster than *growth rings* especially in *mature* or *senescent* trees (NSW Dept. Environment & Conservation 2005, p. 76). Note: this concept appears contrary to CODIT Wall 4 which is laid down at a constant rate in seasonal *growth rings* except where a *wound margin* is stimulated by *tropic* responses to *loading,* e.g. *torsion, compression* or *tension*.

Accelerated growth callus The concept where *xylem* develops by *accelerated growth* at the *wound margins* of wounded *sapwood* (NSW Dept. Environment & Conservation 2005, p. 76). Note: this concept appears confused between the initially formed undifferentiated *wood* as *callus wood* and the later forming differentiated wood as *wound wood*. See also *Wound margin, Wound wood, Callus wood* and *Compartmentalisation of decay in trees (CODIT)*.

Accessory bud Lateral bud associated with a main bud such as in a leaf *axil* and usually develops after damage to the main bud.

Accessory roots The *lateral roots* as with palms, which develop from the base of the trunk different to those arising from the *radicle* of the seed root system.

Accessory trunk *Aerial root* mass differentiated once it reaches the ground forming a vertical woody structure to support a trunk or branch, a *columnar root* or *stilt root*. Here the supported branch is able to extend further and tends to horizontal with the overall *crown spread* covering a considerable area, e.g. *Ficus columnaris*.

Acid sulphate soil Very acidic soil layers or horizons – the result of aeration of soil materials abundant in iron sulphides, mainly pyrite and the result of drying and aeration of previously saturated anaerobic soil material. Usually with a pH of <4.0 when measured in dry seasons conditions and overlying potential acid sulphate soils or soils with more than 0.05% oxidisable sulphur.

Acoustic resonance Auditory reverberation within an object and the air after the object has been struck. See also *Sounding*.

Active maintenance See *Planned maintenance.*

Acute branch crotch A branch crotch where the angle on the inner side of the union is less than <90°. See also *Obtuse branch crotch.*

Acutely convergent A branch growing in a direction towards its point of attachment where the angle in the crotch is less than <90°.

Acutely divergent A branch growing in a direction away from its point of attachment where the angle in the crotch is less than <90°.

Adaptive growth See *Adaptive wood.*

Adaptive wood Additional load-bearing wood formed in response to mechanical stresses and gravitational force upon the *vascular cambium* to provide a uniform distribution of loading. Examples are *Ribs*, *Round-edged rib* or *Sharp-edged rib* and *Buttresses*. See also *Reaction wood*, *Compression wood* and *Tension wood.*

Adaxial The side of a leaf, branch or other organ which anatomically faces towards the *axis* of the parent shoot (i.e. usually the upperside) (Lonsdale 1999, p. 309).

Adventitious A bud arising from points other than terminals or axils, e.g. from a root or at an *internodal* region (Harris *et al.* 2004, p. 15).

Adventitious bud A bud formed within the *cambial zone* and *callus wood* after wounding (Shigo 1989a, p. 134).

Adventitious shoot A branch from a bud arising in an unusual location, e.g. *sucker.*

Adventitious root mass Palms and other monocotyledons may form masses of *fine roots* or *adventitious roots* as *primary growth* where the *radicle* is replaced by branching many times and this may extend above ground and be evident at the base of the trunk. The extent of the root mass above ground may be extensive in some palms and increases with age giving the appearance of lifting the trunk, e.g. *Phoenix canariensis.*

Adventitious roots 1. Roots that may arise in an unusual location and may develop a structural function, e.g. (a) from a branch into a *pocket crotch* where accumulated *leaf litter* and moisture has formed humus, (b) into the *hollow* section of a branch or trunk often where *humus* has accumulated, (c) *aerial roots, column roots, fibrous roots.* 2. Roots that may arise where the *radicle* is replaced by lateral branching many times as with *palms* or grasses.

Advocate An individual or party acting as a representative in support of an issue.

Aerial inspection Assessment of the crown of a tree by climbing within the *crown* or by the use of an *elevating work platform*, often to examine a particular *defect*, e.g. *cavity* or *hollow.* See also *Visual tree assessment* (VTA).

Aerial roots Adventitious roots growing into the air from any above ground part of a tree which may eventually develop a structural function.

Aerobic Living in the presence of oxygen or conditions where oxygen is freely available.

Aerophore See *Pneumatophore.*

Aerotropism Growth direction of a plant or plant part responding to the presence of air.

Age Most trees have a stable biomass for the major proportion of their life. The estimation of the age of a tree is based on the knowledge of the expected lifespan of the taxa *in situ* divided into three distinct stages of measurable biomass, when the exact age of the tree from its date of cultivation or planting is unknown and can be categorised as *young, mature* and *over-mature* (British Standards 1991, p. 13; Harris *et al.* 2004, p. 262).

Air gaps Barriers to root growth formed by load-bearing stone matrices with large voids, e.g. broken bricks, gravel >20 mm; not filled in, that drain well allowing the air to desiccate new roots (Coder 1998, p. 62), e.g. under pavements and behind walls.

Air knife A pneumatic device that uses a fine stream of compressed air with sufficient pressure to displace soil or cut roots. At lower pressure, soil may be displaced allowing woody roots to be exposed for examination or *root mapping*. See also *Water knife*.

Air spade See *Air knife*.

Allelopathy The release of chemicals from a plant that are detrimental to other plants to inhibit the growth of nearby plants, including its own progeny, to reduce competition, e.g. from *Pinus* spp., *Casuarina* spp., *Cinnamomum camphora*, *Eucalyptus* spp.

Alternation of generation Staged replacement planting of an avenue or stand of trees, e.g. by a roadside or park, where new plantings are setback from the originals, ultimately to replace them in a similar configuration. Such an undertaking may be utilised for road widening or to reduce the hazard of vehicular collisions with trees or reduce the impact of removing prominent senescent trees.

Amendment The changing of a *planning provision* controlling land use and development.

Amenity A positive element or elements which contribute to the overall character and pleasantness of an area, e.g. trees, old buildings their *curtilage* and interrelated elements within the environment.

Amenity tree A tree with recreational, functional, environmental, ecological, social, health or aesthetic value rather than for production purposes (Australian Standard 2007, p. 5), and may be synonymous with *shade tree* in the USA.

Anaerobic Living in the absence of oxygen, e.g. anaerobic bacteria.

Anastomosing A plant part subject to the process of *anastomosis,* e.g. roots and stems.

Anastomosis Cross-linking of branching parts, e.g. roots or branches in woody plants where such growth usually forms a *graft,* e.g. 1. Aerial roots of *Ficus* spp., especially in a parasitic situation where a strangler fig germinates in the crown of a host sending *aerial roots* to the ground and around the trunk of the host eventually encasing it, constricting its growth as they enlarge and merge forming a *hollow* trunk structure killing the host. 2. Aerial roots on *Ficus* spp., differentiating to form *column roots* once they reach the ground, providing support for lateral branches. 3. Artificially where *Ficus* spp., are plaited together when young to form a standard potted specimen. 4. Artificially when *pleaching* to form an arbour of intertwined branches.

Anchorage Where sufficient cohesion between roots and soil exists for a tree to maintain *stability.* Stimulus for such root growth results from the flow of forces through the branches along the trunk to the root system.

Anchor roots See *Structural roots.*

Angiosperms Plants where the ovule is fully enclosed within the fruit i.e. container seed. These are the flowering plants and generally referred to as hardwood trees although some have soft non-durable wood. See also *Gymnosperms.*

Anion A negatively charged *ion* (Handreck & Black 2002, p. 16).

Annual growth rings See *Growth rings.*

Annual ring See *Growth rings.*

Annular Ring scars prominent on the trunk of some palms after leaf fall, e.g. *Archontophoenix* spp. (Jones 1996, p. 266).

Anti-transpirant Substance applied to plants to block stomata temporarily to reduce moisture loss by preventing *transpiration*. Often used when *transplanting* trees.

Apedal *Soil horizons* formed without *peds* as part of the *soil structure.*

Apex The tip or furthest point, or the highest point, or the distal end of a leaf, stem or wound.

Aphototropic Growth direction taken showing no response to the stimulus of light, e.g. roots.

Apical Forming at the *apex*.

Apical bud A bud formed at the apex – usually at the end of a branch and is terminal, dominant at the highest point on a tree at the tip of a branch or stem and at the ends of lateral branches.

Apical dominance Suppression of the development of lateral buds by plant growth regulator chemicals produced in the *apical meristem* to promote stem elongation in preference to branching, further stimulated by competition for space and light.

Apical meristem Meristematic tissue at the tips of roots or stems giving rise to primary tissues that are responsible for increasing length rather than girth of the axis. See also *Apical bud* and *Apical dominance*.

Apoplast Interconnected non-living portion of plant tissue including spaces within and between cells and cell walls.

Applicant 1. The property owner or their authorised agent that lodges an application for *development* works requiring approval from a *consent authority*. 2. Individual or party petitioning a court to hear a matter of disputation to seek resolution.

Appropriate tree management The management of trees as a resource based on sound professional judgement and a competent understanding of what tree to plant where and when, or when to remove or retain a tree. Examples: 1. The planting or retention of a tree in a position that causes minimal or no conflict with people or property or disturbance of the built environment, or services or infrastructure, due to such a decision having been founded upon a competent knowledge of the characteristics of the tree's growth pattern and ultimate dimensions above and below ground at maturity, and the suitability of the space available into which it will develop. 2. The removal of a tree that will grow to be in conflict with the constraints of its growing *environment* either above or below ground at its ultimate dimensions at maturity, and especially where replanting could be undertaken with an advanced specimen of a species of more suitable growth characteristics and mature dimensions. 3. The removal of a vigorous tree in a *poor condition*, in a prominent position where its potential failure in full or part poses a risk of hazard to the safety of people, or damage to property. See also *Inappropriate tree management* and *Tree management.*

Arbour A walkway covered by the growth of vines or the branches of trees usually cultivated for that purpose.

Arbor Day A day set aside for planting trees. Julius Sterling Morton (1832–1902) introduced the concept on 4 January, 1872, in Nebraska, USA, to promote the benefits of tree planting in areas with no trees or where trees had been removed. The first tree planting day was held on 10 April, 1872 and the day itself was observed after state proclamation on 10 April, 1874. In 1885, Arbor Day was named a legal holiday in Nebraska and 22 April, Morton's birthday, was selected as the date for its permanent observance. The tradition soon spread around the world and is celebrated in most countries at different times of the year. Table 1 details some Arbor Day celebrations around the world (The National Arbor Day Foundation 2005).

Arboreal Living in or connected with trees.

Arborescent Developing to appear like a tree, especially with branching form.

Arboretum An area planted with a variety of trees, woody shrubs and vines for purposes of research, conservation and display.

Arboricultural Pertaining to *Arboriculture.*

Arboricultural consultant See *Consulting arboriculturist.*

Arboriculture The science and culture of the growth, planning, management, care and maintenance of trees primarily for amenity and utility purposes. See also *Tree management, Tree preservation* and *Urban forestry.*

Arboriculturist 1. An individual with competence in the science of Arboriculture with skills specialised in practices for the planning and management of trees, usually in urban environments, primarily for *amenity* and utility purposes. 2. Synonymous with *Arborist*, especially in the USA.

Table 1 Details of some Arbor Day celebrations around the world (The National Arbor Day Foundation 2005). Dates in individual countries may change over time.

COUNTRY OR STATE	TITLE OF CELEBRATION	DATE CELEBRATED
Australia Western Australia Northern Territory Queensland South Australia New South Wales Victoria Australian Capital Territory Tasmania	Arbor Day	 June Nov May June Monday of last week in July Arbor Week last week in July Last week in June 27 July Oct
Brazil Araras, Sao Paulo	Festa das 'Arvores	June 7
Canada: Ontario Nova Scotia	Arbor Day	Last Friday in April to Sunday in May. First full week in May
China	Arbor Day	March
Commonwealth of the Northern Mariana	Arbor Day	Oct 1
Germany	Arbor Day, Tag des Baumes	April 25
Guam	Arbor Day	first Tuesday of October
Iceland	Students' Afforestation Day	Not Listed
India	National Festival of Tree Planting	Not Listed
Israel	New Year of the Trees	15th day of the Hebrew month of Shevat
Japan	Greenery Day or Greening Week, (midori noni)	late April
Korea	Tree-Loving Week	early April
Mexico	Dia del Arboles (Day of the Trees)	a day in July
New Zealand	Arbor Day (Also World Environment Day)	June 5
Puerto Rico	Arbor Day	last Friday in September
United Kingdom	National Tree Week National Tree Dressing Day	November first weekend in December
United States of America	National Arbor Day	Last Friday in April

Arborist 1. An individual with competence to cultivate, care and maintain trees for *amenity* or utility purposes. 2. Synonymous with *arboriculturist*, especially in the USA.

Arborsonic Decay Detector® See *Sonic detectors*.

Arborvitae See *Arbor vitae*.

Arbor vitae 'Tree of Life', a reference to the genus *Thuja*, the bark from which was once used by sailors to make a tea rich in vitamin C to prevent scurvy (Spencer 1995, p. 212).

Arbor Week In some countries this is a week-long celebration as an extension of *Arbor Day*.

Architect An expert in the consultation design and documentation of buildings and supervision of their construction.

Architecture Description of branching patterns in the crown or root system (Lonsdale 1999, p. 310).

Area within dripline See *Crown projection*.

Arrangement of first order branches within a crown The pattern formed by the first order branches at the point of their attachment to the trunk. (See Figure 1.)

Ascending hollow A *hollow* that develops upwards in a trunk or branch usually in a *distal* direction. See also *Descending hollow* and *Hollow*.

Aspect ratio The diameter of a branch compared to the diameter of the trunk. The diameter of the branch measured at its base is divided by the diameter of the trunk measured immediately above the branch bark ridge and the branch diameter measured immediately above the branch bark ridge and branch collar (Gilman 2003, pp. 291–292).

Asserted dominance 1. In a grafted tree, branches arising from the *understock* below the *graft* union become more vigorous than the *scion* rendering it inferior. 2. Branches previously inferior or codominant as dual-leader branches or a *lateral* becoming erect or corrected to upright through *photo-*

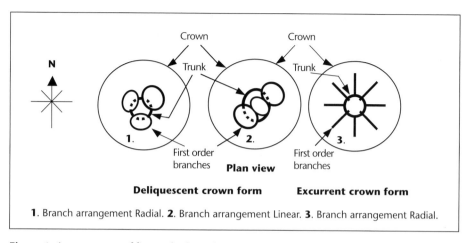

Figure 1 Arrangement of first order branches within a crown.

tropism and *gravitropism* and negative-geotropism after the death or damage of the superior *first order structural branch* or *apical meristem*. 3. The return to *apical dominance* by one or more epicormic shoots as an *elite* after an *episode* causing the destruction of the crown or trunk, including shooting from a lignotuber, e.g. as a result of fire, lopping, drought, severe insect damage, storm damage, poisoning, root severance etc.

Assessment A criterion to estimate or determine the value or magnitude of a tree for its monetary or intrinsic worth to assist with its management, considering many different attributes, e.g. *age*, *amenity* value, significance, *condition*, *form*, *viability*, safety, *vigour*, *symmetry*, *defects*.

Assessor See *Commissioner*.

Asymmetrical Imbalance within a crown, where there is an uneven distribution of branches and the foliage *crown* or *root plate* around the vertical *axis* of the trunk. This may be due to *crown form codominant* or *crown form suppressed* as a result of natural restrictions, e.g. from buildings, or from competition for space and light with other trees, or from exposure to wind, or artificially caused by pruning for clearance of roads, buildings or power lines. An example of an expression of this may be, crown asymmetrical, bias to west. See also *Symmetrical* and *Symmetry*.

Asymmetrical cavity A *cavity* formed with a generally uneven development from the *axis* towards one or more sides of a stem or root. See also *Symmetrical cavity.*

Asymmetrical decay An area of *decay* formed with a generally uneven development from the *axis* towards one or more sides of a stem or root. See also *Symmetrical decay.*

At grade See *Grade.*

Attached broken branch/frond A live or dead branch or palm frond, that has snapped or fractured damaging its wood, destroying *structural integrity* at its point of connection, or has been compartmentalised by abscission (the frond), but remains joined to the tree at this point. (See Figure 2.)

Atypical Having an appearance that does not conform to that of others in a taxonomic group. In a tree this may also be growth that is not representative of crown form, habit and type or behaviour expected to occur naturally. See also *Misshapen* and *Typical.*

Australian Height Datum (AHD) The datum (adopted by the National Mapping Council of Australia) to which all vertical control for mapping is to be referred in Australia.

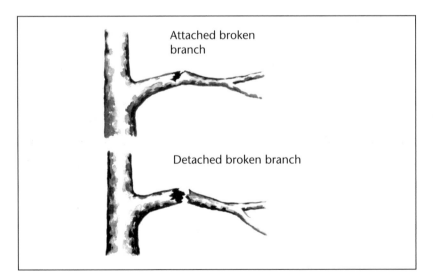

Figure 2 Difference between an attached and detached broken branch/frond.

Auto-graft See *Grafted branches.*

Autotroph Organism that in order to survive synthesises *organic* matter from *inorganic* nutrients and an energy source (being light for most plants). See also *Heterotroph.*

Available water The maximum amount of water roots can obtain from soils between *field capacity* and *permanent wilting point* (Handreck & Black 2002, p. 74).

Axial Root or shoot growth along or parallel to the *axis* of a structure (Lonsdale 1999, p. 310).

Axial pressure See *Axial stress.*

Axial stress Downward loading force exerted by the mass of the *crown* along the *axis* of the trunk.

Axil Angle situated between the *adaxial* side of a leaf, branch or other organ and the *axis* on which it is borne (Lonsdale 1999, p. 310).

Axillary Forming at an *axil.*

Axillary bud A bud in the angle between a leaf and stem, or between two stems. Such buds are often dormant and may remain undeveloped until the *apical bud* is damaged, removed or suppressed.

Axillary dominance The dominance asserted by axillary growth after damage, removal or suppression of the *apical bud.* This is also known as secondary dominance.

Axiom of uniform stress The principle that a tree is mechanically optimised growing only sufficient wood for support and loading. As a result, no area is under-loaded to breaking point or over-loaded with excess material (Mattheck & Breloer 1994, pp. 12–13).

Axis (*pl.* axes) A main stem, trunk or root, or the direction in which a structure is orientated (Lonsdale 1999, p. 310), or the growth of plant parts from its centre, e.g. *crown spread orientation.*

Backflash Movement of herbicide from one tree to another through a *root graft* (Thomas 2000, p. 98).

Baiting *Directed growth* of roots away from existing *infrastructure* enticing them to develop in an alternate location or direction by providing essential resources in good soil (Coder 1998, p. 65). See also *Channelling*.

Ball and burlap See *Balled-in-burlap*.

Balled-in-burlap A method of containing or transporting the *root ball* of advanced trees typically grown in a field *environment* where trees are mass produced for use in the landscape. The *root ball* is wrapped in hessian or gunny fabric for *transplanting* and remains attached to the root ball when placed into the new planting hole. The fabric around the root ball is retained at the time of planting, buried by back filling and allowed to decompose.

Balled and burlapped See *Balled-in-burlap*.

Banana crack See *Subsidence crack*.

Bare root See *Bare root stock*.

Bare root stock Preparation of a plant for *transplanting* where the soil is removed from the *root plate* or root ball, and only a few lower *orders of roots* are retained.

Bark The protective coating of tissues to the outer side of the *vascular cambium* (Shigo 1986, p. 5) divided into the *inner bark – phloem* and *outer bark – phellem*. The living cells are added incrementally to trunk, branches and roots from *phloem* also called *phellogen*. Bark may persist and build up over many seasons of growth, e.g. *Quercus suber*, or be *decorticated* within each successive growth period, e.g. *Corymbia citriodora* (syn. *Eucalyptus citriodora*).

Bark buckling See *Congested bark*.

Barrier zone Initial protection boundary formed after wounding by still living cambium usually providing a strong barrier to micro-organisms but structurally weak (Shigo 1986, pp. 6–7, 152).

Basal *Proximal* end of the trunk or branch, e.g. trunk *wound* extending to the ground is a *basal wound*, or as *epicormic shoots* arising from a *lignotuber*.

Basal bell fracture The failure of a *bottle butt* as the result of a mixture of *shear failure* and *delamination* from transverse stresses, and dependent on remaining stem wall thickness *shell buckling* or *hose pipe kinking* may result (Mattheck & Breloer 1994, pp. 44–45).

Basal flare Swelling at the root crown usually uniform around the base of the trunk involving tissue from the trunk and root crown. Here *first order roots* may not be evident at the root crown, e.g. *Lophostemon confertus*.

Basal swelling See *Bottle butt*.

Basal trunk wound A wound on the trunk extending to the *root crown* where the base of the wound is open at the ground and usually truncated. Dependent upon the width of its base such a wound may not become *occluded*. (See Figure 3.)

Bast (Scott 1912, p. 52) See *Phloem*.

Batter A constructed or naturally formed slope often to denote the intensity of angle.

Beam A usually elongated horizontal support for a built structure used to span across a void usually from one *pier* to another.

Bedrock Rock strata located below soil layers, but may extend above ground.

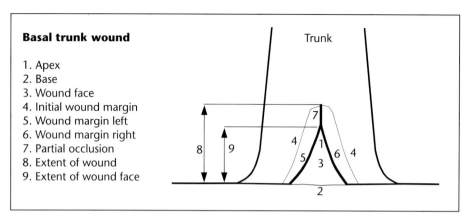

Figure 3 Basal trunk wound.

Benching Relatively level strips of earth or rock in a broad step-like formation breaking the continuity of a slope usually for reasons of safety. Depending on soil and rock type a commonly used ratio for vertical to horizontal cutting is 2:1.

Benchmark Permanent point of reference made by a surveyor on a site or area which all data is referred to.

Bifacial vascular cambium See *Secondary thickening.*

Bifurcate Roots or branches divided at one end into two parts. See also *Fork.*

Bifurcation The process of the division of roots or branches at one end into two parts. See also *Fork.*

Biodiversity The variety of life forms, plants, animals and micro-organisms, usually considered at three levels, genetic diversity, species diversity and ecosystem diversity (Willoughby City Council 2008).

Biological control A method of reducing the number of a pest species or disease-causing organisms by encouraging predatory organisms or diseases to attack it.

Biological control agent An organism which, through artificial introduction or augmentation of its population, helps to protect a crop, an ecosystem or an individual plant or animal against a pest, pathogen or *decay* organism (Lonsdale 1999, p. 310).

Biomass Mass of living matter (plant and/or animal and/or microbial), in a given area at a particular time.

Biotic Living, e.g. fauna, flora, fungi, viruses, bacteria, humans. See also *Abiotic* and *Abiotic factors*.

Blaze A wound cut into a tree usually to the *sapwood* and sometimes extending to *heartwood* to create a marker point, e.g. by a surveyor, the *wound face* may be further incised or painted to denote additional information.

Bleeding See *Exudate*.

Blowholes A slightly raised area of *wound wood* tissue that forms after a longitudinal slit is cut along the trunk from the inner termite nest to allow the alates (winged reproductives) to leave to form a new colony. After the departure of the alates the holes are sealed by workers assisting the growth of *callus wood* and then *wound wood* until *occlusion* occurs, leaving a small narrow scar characteristically 80–100 × <5 mm indicative of the termite nest within the tree.

Body language See *Body language of trees*.

Body language of trees Apparent *typical* growth patterns in a tree or *atypical* growth patterns resulting from deformation of growth in response to loading by mechanical stresses (Lonsdale 1999, p. 311).

Bole See *Trunk*.

Bottle butt Uncharacteristic bulging stem growth at the base of the trunk due to altered *stress* in this region, often associated with *decay* (Lonsdale 1999, p. 311).

Bow Bend in a trunk or branch formed over time due to gradual increases in *loading*. See also *Butt sweep*.

Bracing Systems of cables and ropes, traditionally using metal wires but generally replaced by polypropylene, used to support and prolong the life of trees in part or full; systems are susceptible to failure due to evident weaknesses in branch unions. Examples of such proprietary systems are Cobra, and Yale nylon webbing.

Bracket See *Bracket fungus.*

Bracket fungus The rigid *sporophore* of some fungus species especially those associated with live trees or the *decay* of wood. Structures comprised of hyphae for the dispersal of spores, often bracket shaped usually protruding from the roots, trunk or branches of a host tree when the fungus matures. The fruiting body may be *ephemeral* or persistent and may last for only one season or persist for many years with the fruiting body growing incrementally larger and continuing to produce new spores. Such fruiting bodies may be solitary or gregarious.

Branch An elongated woody structure arising initially from the trunk to support leaves, flowers, fruit and the development of other branches. A branch may itself fork and continue to divide many times as successive *orders of branches* with the length and taper decreasing incrementally to the *outer extremity* of the *crown*. These may develop initially as a gradually tapering continuation of the *trunk* with minimal division as in a *young* tree or a tree of *excurrent habit*, or in a *sapling*, or may arise where the trunk terminates at or some distance from the *root crown*, dividing into *first order branches* to form and support the *foliage crown*. In an *acaulescent* tree, branches arise at or near the *root crown*. Similarly branches may arise from a *sprout mass* from damaged *roots*, *branches* or *trunk*.

Branch attachment See *Branch union.*

Branch bark inclusion See *Included bark.*

Branch bark ridge *Extruded bark* forming a convex protrusion or striation or series of ripples in the *crotch* of the *branch union*. See also *Included bark.*

Branch collar The swollen ring of growth formed around the base of a branch by the successive layers of each *growth increment* of the branch and the supporting branch or trunk to which it is connected growing and intertwining around its edges (Shigo 2004, p. 5).

Branch core After a branch fails or is removed, this is the remaining branch section within the connecting branch or trunk walled off by *compartmentalisation* (Shigo 1989a, p. 23). See also *Branch tail.*

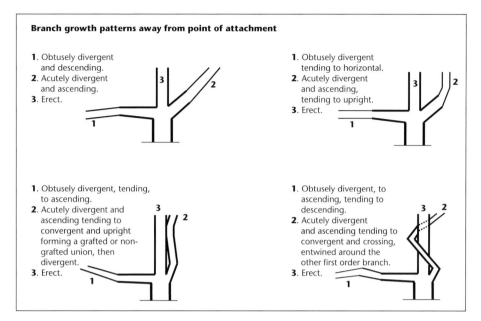

Branch growth patterns away from point of attachment

1. Obtusely divergent and descending.
2. Acutely divergent and ascending.
3. Erect.

1. Obtusely divergent tending to horizontal.
2. Acutely divergent and ascending, tending to upright.
3. Erect.

1. Obtusely divergent, tending, to ascending.
2. Acutely divergent and ascending tending to convergent and upright forming a grafted or non-grafted union, then divergent.
3. Erect.

1. Obtusely divergent, to ascending, tending to descending.
2. Acutely divergent and ascending tending to convergent and crossing, entwined around the other first order branch.
3. Erect.

Figure 4 Examples of branch growth away from the point of attachment.

Branch creep The irreversible *deformation* of wood by the *loading* force of its own weight heightened by high temperatures (Lonsdale 1999, p. 30), e.g. on a long branch tending to horizontal.

Branch defence zone See *Branch protection zone*.

Branch failure The structural collapse of a branch that is physically weakened by wounding or from the actions of pests and diseases, or overcome by loading forces in excess of its load-bearing capacity.

Branch growth away from the point of attachment This is the orientation of branch growth from its point of attachment and the angles in the crotches being *divergent, convergent* or *erect, acute* and *obtuse.* (See Figure 4.)

Branchlet Lowest order of branching that differs to a *twig* often by the presence of persistent foliage in *Gymnosperms*, e.g. *Araucariaceae*.

Branch junction See *Branch union*.

Branch protection zone Zones formed as a chemical boundary at or near the proximal end of branches to kill or repel the spread of micro-organisms (Shigo 1989a, p. 239).

Branch-shedding collar A *branch collar* continuing to develop around the remains of a dead branch.

Branch socket See *Tear out wound.*

Branch structure within crown The arrangement of first order branches from the trunk. This may be considered as *deliquescent* and *excurrent* with some structures intergrading between the two. (See Figure 5.)

Branch stub A section of branch remaining beyond the branch collar and usually evident after a *lopping* episode or as a result of *branch failure*, or *natural pruning.*

Branch subsidence Gradual downward bending of branches, usually seen with long branches tending to horizontal succumbing to loading stresses under their own weight. See also *Subsidence crack* and *Sudden branch drop.*

Branch tail The tapering underside of a branch at its proximal end where its fibres intertwine to provide some structural support with the fibres of the branch or trunk where it is attached and new layers of such growth are added by each successive *growth increment*, however, the *branch collar* forms the greater majority of strength of the branch union (Shigo 1989a, pp. 215–217). See also *Branch core.*

Branch tear See *Branch tear out.*

Branch tear out Dislodging of a branch from its point of attachment where it is torn away from the *branch collar* snapping the *branch tail* causing a *laceration*, usually to the underside of the *branch union* of the branch or trunk to which it was attached forming a *tear out wound.*

Branch tear wound See *Tear out wound.*

Branch trace See *Branch tail.*

Branch:Trunk diameter ratio See *Aspect ratio.*

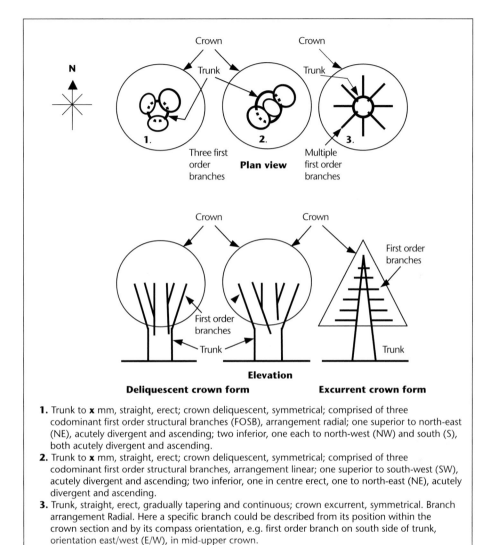

Figure 5 Branch orientation. The compass direction of a first order branch as it grows away from the trunk or from the point of attachment for a lower order branch.

Branch union Place of common juncture for two or more branches where each annual *growth increment* overlaps around and below the union and successive layers are embedded further into the trunk or supporting branch. At this intersection a distinct ring may form on the larger supporting branch around the smaller branch forming a *branch collar*, or may be absent if the two or more branches were codominant and of equal size having arisen at the same time and had grown at similar rates in competition with each other.

Branch-wood Wood formed in a *branch*.

Branch-wood tail See *Branch tail*.

Breathing roots See *Pneumatophore*.

Bridge footing Isolated footings located on both sides of a tree's root to be protected, with support for the built structure provided by *beams* or *lintels* spanning between the footings.

Brown rot Decomposition by fungi of the carbohydrate component of wood, being the carbohydrates *cellulose* and *hemicellulose*, leaving much of the brown coloured *lignin* making the wood appear brown. In its later stages this rot generally causes fractures across the grain breaking the wood into small or large coloured cubes, a distinctive identification tool for this fungus. This rot is usually caused by fungi from the subdivisions *Hymenomycetes* of the subdivision *Basidiomycotina* (Manion 1991, pp. 227–228). This *decay* affects the compressive strength of wood.

Buckling Irreversible deformation of a structure subjected to *compression* from a bending load (Lonsdale 1999, p. 311).

Bud An active or dormant embryonic outgrowth on a branch/stem consisting of a short stem terminated by a meristem and, in foliage buds, numerous leaf primordium, leaf buttresses, and young rolled or folded leaves and may be enclosed by protective scales. Buds may give rise to either, leaves, stems or flowers.

Bud burl See *Pollard head*.

Bud scales See *Cataphylls*.

Bud scale scar Scar on branches from the abscission of bud scales that denote the cessation of growth from the previous season's *extension growth* at the distal end of a stem and the commencement of new growth in a successive growing season.

Building envelope Total surface area of ground that is or will be covered by a building.

Building footprint See *Building envelope.*

Bulge wood See *Ring swelling.*

Bulk density The weight of a unit of dried soil including the solid and *pore* volumes (Craul 1992, p. 15).

Bulk excavation See *Soil cutting.*

Bulking See *Over excavation.*

Burl Woody protrusion found usually on the trunk, often rounded or hemispherical, may reach substantial dimensions possibly formed by a cambial disruption that may be ongoing as a bud *fasciation* in a massive cluster. They may also form where buds are removed annually from a trunk, or at the graft intersection of an *understock* and *scion* but here it is likely to be formed from callus and wound wood.

Burr See *Burl.*

Bushfire hazard reduction A reduction or modification of fuel by burning or by chemical, mechanical or manual means (Willoughby City Council 2008).

Bushland Land supporting *remnant vegetation* or, where disturbed, vegetation is still representative of the floristics and structure of its original unaltered state.

Bushland vegetation Plants which are *remnant vegetation* of the natural vegetation of the land where they are situated or, if altered, are still representative of the structure and floristics of the natural vegetation (Willoughby City Council 2008).

Bush regeneration Rehabilitation of *bushland* from a weed infested or otherwise degraded plant community to vegetation representative of the floristics and structure of its original unaltered state where natural regrowth can occur (Buchanan 1989, pp. 6–7).

Bush regenerator Individual specialising in and undertaking *bush regeneration.*

Butt The lowest part of a trunk.

Buttress A flange of *adaptive wood* occurring at a junction of a trunk and root or trunk and branch in response to addition loading.

Buttress root A flange of *adaptive wood* as an upright extension of the *first order roots* and the trunk adding to the *stability* of many rainforest taxa, and often on tall trees. The flange tapers up the trunk and out along the first order root where it may extend several metres from the trunk. It may extend to branches and branch collars on trees with short trunks.

Buttress wood A structural flange formed by loading at the junction of a trunk or branch. See also *Buttress* and *Buttress root*.

Buttress zone Area at the base of the trunk where *buttress roots* form or the distance from the trunk to which they extend.

Butt rot A form of *decay* in standing trees, which primarily affects the roots and *buttress zone* but may also extend up the trunk (Lonsdale 1999, p. 320).

Butt sweep A forestry term for a substantial curve in the trunk of a tree near the ground. Butt sweep occurs when a tree is partially windthrown when young, but then stabilises itself and straightens and is *self-correcting* due to *reaction wood* (Stephen 2003).

Butt swell See *Basal flare*.

Cables See *Bracing.*

Callus wood Undifferentiated and unlignified wood that forms initially after wounding around the margins of a wound separating damaged existing wood from the later forming lignified wood or *wound wood.*

Calyptra *Parenchyma* cells at the *root tip*, shaped like a hood that protects it from abrasion as it extends through the soil.

Cambial zone A region of lateral meristem comprised of conductive cells which gives rise to *secondary xylem* and *secondary phloem.*

Cambium See *Vascular cambium.*

Canker A *wound* created by repeated localised killing of the *vascular cambium* and bark by wood *decay* fungi and bacteria usually marked by concentric disfiguration. The wound may appear as a depression as each successive *growth increment* develops around the *lesion* forming a *wound margin* (Shigo 1991, p. 140; Keane *et al.* 2000, p. 332).

Canker rot See *Canker.*

Canopy 1. Of multiple trees, the convergence, or merging in full or part, of the crowns of two or more trees due to their proximity, or where competition for light and space available in a forest environment is limited as each tree develops forming a continuous layer of foliage. 2. Used as a plural for *crown*. 3. Sometimes synonymously used for *crown* (USA).

Canopy cleaning See *Crown maintenance.*

Canopy closure The converging of crowns of adjacent trees, so that little direct sunlight reaches the ground.

Canopy cover The amount of an area of land covered by the lateral spread of the tree *canopy* when viewed from above that land, e.g. as determined by aerial photography and usually expressed as a percentage of surface of land area.

Canopy roots Adventitious *fine roots* formed within the crown of some rainforest trees utilising moist *humic soil* in branch crotches and hollows or along or around the trunk, which benefit from nutrients leached from the crown. Unlike other *aerial roots* these roots do not try to grow to the ground (Thomas 2000, pp. 88–89).

Cantilever A projecting built structure attached or supported at two points at only one end generally allowing a third of the supported length to extend over an area without disturbance, e.g. over tree roots.

Carbon sequestration Absorption and storage of carbon by plants, especially trees and algae.

Carved tree A pattern cut into the wood of a living tree, usually on the trunk as a marker or for ceremonial purposes by indigenous people, or sometimes for the purposes of art. See also *Scar tree.*

Cataphylls Modified leaves forming scales around buds, often resinous, resisting desiccation to protect dormant buds on trees over winter.

Carpophore See *Sporophore.*

Cation A positively charged *ion* (Handreck & Black 2002, p. 16).

Cation Exchange Capacity (CEC) The measurement of the exchange of positively charged nutrients in solution with the negatively charged surfaces of soil particles (*colloids*) freeing the exchanged *anions* for absorption by roots (Handreck & Black 2002, p. 48).

Caulescent Tree grows to form a *trunk*. See also *Acaulescent*.

Cauliflorous Trees where the flowers or inflorescences arise from the trunk or branches, not from *apical buds* or *axillary buds*.

Cauliflory Flowers or inflorescences arising from the trunk or branches.

Cavity A usually shallow void often localised initiated by a *wound* and subsequent *decay* within the trunk, branches or roots, or beneath bark, and may be enclosed or have one or more opening. See also *Hollow*.

Cellulose A structural substance in cell walls comprised of long chains of glucose molecules unable to be metabolised by the plant after absorption into the cell wall.

Centre of trunk (COT) Estimated position of the middle of a trunk from where a measurement is taken, e.g. the boundary was 2.5 m from COT. See also *Edge of trunk*.

Central root system See *Structural roots*.

Channel See *Gutter*.

Channelling *Directed growth* of roots away from existing *infrastructure* or where space is limited enticing them to develop in an alternate location or direction by physically guiding or constraining their development, e.g. near surface trenches, channels, layers, tunnels surrounded by root barriers (Coder 1998, p. 65). See also *Baiting*.

Chemically inhibiting See *Root barrier* and *Inhibiting*.

Chemotropism A directional growth movement of a plant or part in response to a chemical stimulus.

Chinese moustache See *Crotch seam*.

Chlorophyll fluorescence meter An electronic device to measure stress in trees determined by changes in the rate of photosynthesis.

Chlorosis A condition in plants resulting from the failure of chlorophyll to develop, usually due to a deficiency of an essential element and evident in leaves as a discolouration ranging from light green, yellow to almost white. See also *Essential elements*, *Macronutrients* and *Micronutrients*.

Chlorotic Yellowing of leaves caused by the absence of chlorophyll.

Circling root The growth of roots that is not radial away from the trunk and curves to encircle the trunk.

Circling root barrier *Deflecting root barrier* usually plastic that encircles the roots to varying depths, e.g. 300 mm, 600 mm, usually installed at the time of planting.

Circumferential compressive stress On straight trees a *compression* force at the surface of a stem exerted longitudinally, evident as narrower spindle shaped *vascular rays* (Mattheck & Breloer 1994, pp. 168, 169).

Cladoptosis The process of shedding twigs by *abscission* (Lonsdale 1999, p. 312).

Climate Average weather patterns in a particular region over a period of time. See also *Microclimate*, *Mesoclimate* and *Macroclimate*.

Climbing roots Any short *adventitious roots* that may develop from the stems of some climbing plants, e.g. *Hedera helix*, to attach the plant to its support.

Clinometer An instrument used to determine angles of slope, adapted to measure tree height.

Clod A *ped* larger than 5 mm diameter (Handreck & Black 2002, p. 52).

Closing over See *Partial occlusion*.

Coarse frass *Frass,* grain-like to the touch with particles greater than >1 mm in diameter. See also *Frass*, *Fine frass* and *Medium frass*.

Code of ethics An agreed charter of accepted moral behaviour and standards in business relations and within a profession.

CODIT Wall 1 Wall 1 prevents the vertical spread of pathogens by blocking the cell above and below. After being wounded, the tree responds in a dynamic way by plugging the vertical vascular system above and below the wound. The conducting elements (vessels in angiosperms and tracheids in gymno-

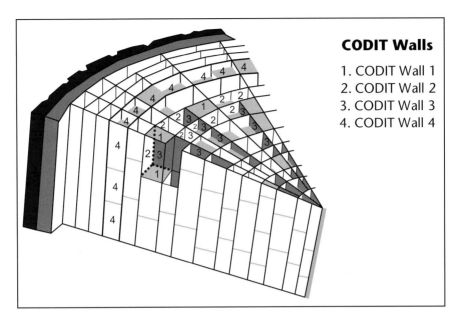

CODIT Walls

1. CODIT Wall 1
2. CODIT Wall 2
3. CODIT Wall 3
4. CODIT Wall 4

Figure 6 Compartmentalisation of decay in trees (CODIT). (Source: USDA 1979)

sperms) are plugged in various ways: tyloses, gum deposits, pit aspera-tions, etc. The plugged elements complete the transverse top and bottom walls of the compartments. Wall 1 is the weakest wall (Shigo 1979). (See Figure 6.)

CODIT Wall 2 Wall 2 uses growth rings to prevent the inward spread of patho-gens. The last cells to form in each growth ring make up the tangential walls of the compartments. These walls are **continuous** around each growth ring except where sheets of *vascular rays* pass through. Wall 2 is the second weakest wall (Shigo 1979). (See Figure 6.)

CODIT Wall 3 Wall 3 uses *vascular rays* to prevent the sideways spread of pathogens. Sheets of *vascular ray* make up the radial walls. These are **discontinuous** walls because they vary greatly in length, thickness, and height. Walls 3 are the strongest walls in the tree at the time of wounding (Shigo 1979). (See Figure 6.)

CODIT Wall 4 Wall 4 grows new cells to conceal the *wound* and restore struc-tural strength. After a tree is wounded, the *vascular cambium* begins to form a new protective wall. The wall is both an anatomical and a chemical wall.

This wall separates the tissue present at the time of wounding from tissue that forms after. It is the strongest of the four walls (Shigo 1979). (See Figure 6.)

Codominant branch Two or more *first order structural branches* or lower order branches of similar dimensions arising from about the same position from a trunk or stem.

Codominant first order structural branches Two or more *first order structural branches* forming a *crown*.

Codominant stem See *Codominant branch*.

Collapse Fall down suddenly usually from *structural failure*, weakness or lack of support.

Collapse and hinge branch failure Branch failure when the crotch of the branch bark union has *included bark* on one side of the *fork* and a weak attachment on the other such as a *sharp-edged rib* or *round-edged rib*, and upon collapse descends but also swings or hinges from the *adaptive wood* of the *rib* as it tears from the point of attachment on one side (Mattheck & Breloer, 1994, pp. 64–65). This is generally a weak branch union.

Collar See *Branch collar*.

Colloids The smallest mineral (clay) and organic (humus) particles in soil able to remain suspended for lengthy periods in pure water (Handreck & Black 2002, p. 44) and considered to have a correspondingly large surface area per unit of mass.

Column boundary layer Interface between a column/s of dysfunctional discoloured or decayed wood and *sound wood*; applicable to a *reaction zone* or a *barrier zone* (Lonsdale 1999, p. 312).

Columnar roots See *Column roots*.

Column roots Aerial root mass differentiated once it reaches the ground *anastomosing* and forming a vertical woody structure to support a trunk or branch, an *accessory trunk* or *stilt root*. Here the supported branch is able to extend further and tends to horizontal with the overall *crown spread* covering a considerable area, e.g. *Ficus columnaris*.

Commissioner An individual empowered to critically examine facts and *evidence* for cases based on merit and make judgments on matters other than law.

Common law Law/s developed over time by the findings of judiciary.

Compartmentalisation See *Compartmentalisation of decay in trees (CODIT)*.

Compartmentalisation of decay in trees (CODIT) (Shigo 1979). A dynamic defence and protection process in trees to resist the spread of pathogens and *decay* organisms using existing and new cells as physical and chemically enhanced barriers (Australian Standard 2007, p. 6) as a system of four walls.

Complying development A *development* proposal satisfying defined planning principles within an *ordinance* of the *consent authority*.

Compression The action of pressure being applied to make something smaller. Evident on the underside of a stem or root usually exhibited as ribs or folds perpendicular to the bending force. This may be evident as buckling bark. See also *Compression wood*.

Compression fork A fork formed where two stems with an *acute branch crotch* grow pressing against each other with *included bark* which becomes *enclosed bark* where the stems flatten at their interface under increasing *compression* from each successive *growth increment*, forming a weak *graft* as a *welded fork* which remains susceptible to *tensile* stress (Mattheck & Breloer 1994, p. 60).

Compression strength Ability of material or a structure to resist failure when subject to *compressive loading* (Lonsdale 1999, p. 312).

Compression stress Loading force from *compression* forces.

Compression wood *Reaction wood* formed by *gymnosperms* as additional wood growth on the under side of a stem opposing a lean, reacting to the loading stimulus to push the stem upwards.

Compressive buttressing Broad *buttressing* of *first order roots* usually below ground on broad trunked trees from temperate regions utilising soil mass above for *anchorage* and to dissipate loading movement away from the *axis* by *compression* on the *leeward* side and resistance to uplift on the *windward* side (Craul 1999, pp. 165–166).

Compressive loading Mechanical loading which exerts a positive pressure; the opposite to *tensile* loading (Lonsdale 1999, p. 312).

Computed tomography Computerised series of *tomography* from scans utilising radioisotopes providing data accurately mapping wood density and moisture content, with further synthesis of each section completing an image to scale of a *stem* showing cavities or areas of different density (Nicolotti & Miglietta 1998, pp. 297–302).

Condition A tree's *crown form* and growth habit, as modified by its *environment* (aspect, suppression by other trees, soils), the *stability* and *viability* of the *root plate*, trunk and structural branches (first (1st) and possibly second (2nd) order branches), including structural defects such as wounds, cavities or hollows, *crooked* trunk or weak trunk/branch junctions and the effects of predation by pests and diseases. These may not be directly connected with *vigour* and it is possible for a tree to be of *normal vigour* but in *poor condition*. Condition can be categorised as *good condition*, *fair condition*, *poor condition* and *dead*.

Cone Female flower structures of some gymnosperms, often woody and becoming the cone fruit after fertilisation. The separate male flower is a *pollen cone*.

Congested bark Localised area of a stem subject to *compression* causing bark to become bunched up as swollen blocks or *buckling* across the stem (Mattheck & Breloer 1994, p. 174). See also *Loosened bark*.

Conk See *Sporophore*.

Conifer Applied generally to *gymnosperms* but more specifically to those trees that develop cone fruit where the seeds are not formed in ovaries.

Consent approved removal Removal of a tree subject to the granting of consent for such works under any law that provides for the removal or protection of trees.

Consent authority The body with the legal power to determine whether or not to grant *development* consent, e.g. Government or a Court.

Consent conditions Requirements placed upon an approved development, either construction or operational based, enacted within legislative and *policy* obligations, e.g. a Government or a Court.

Consequential removal Removal of a tree as a result of increased exposure following changes to its growing *environment* above or below ground where its retention may render it vulnerable to failure in full or part posing a safety risk.

Conservation area An area designated by a *consent authority* under planning *legislation* as possessing special architectural or historical interest whose characteristics are considered worthy of preservation.

Constricting root barrier See *Root barrier* and *Trapping*.

Consulting arboriculturist A professional arboriculturist in a private practice providing a broad range of information and report services to clients regarding trees, usually in urban environments, including tree care, hazard assessment, scientific testing, research, planning, inventories, maintenance and management, building development, human issues, and specialist Arboricultural advice and multi-discipline support to other professions. Synonymous with *Consulting arborist*, especially in the USA.

Consulting arborist See *Consulting arboriculturist*.

Contact stress The sharing of stress *loading* at a point where a tree grows against another tree or structure, further distributing the loading as the surface areas touching increase (Hayes 2001, p. 21).

Contour Prescribed increment depicting the natural shape of land between *contour lines*.

Contour line A line joining points of equal height above sea level.

Convergent branch The direction taken by a branch as it grows towards another. See also *Divergent branch*.

Convex branch bark ridge See *Branch bark ridge*.

Convex stem bark ridge See *Branch bark ridge*.

Coppice 1. The mass of *epicormic shoots* arising after *coppicing*. 2. A dense stand of small trees regularly pruned back to stimulate regrowth.

Coppicing Cutting a tree near to ground level to encourage the development of *epicormic shoots* to produce multiple first order branches in response to the wounding stimulus. Destruction of the *crown* may also stimulate such a response in some taxa, e.g. in some eucalypts after fire with *epicormic shoots* arising from a *lignotuber* (Boland *et al.* 2006, p. 686).

Copse A small area of trees or shrubs growing together. See also *Coppice*.

Cork cambium See *Phellogen*.

Corrective pruning See *Remedial pruning*.

Cortex The tissue including the *endodermis* between the *epidermis* and the *stele* in a stem or root.

Cost analysis A comparison of costs and benefits between different potential courses of action, e.g. tree *removal* and replacement compared with *managed decline*.

Cotyledon The first or seed leaves of plants. See also *Dicotyledon* and *Monocotyledon*.

Council officer An individual employed by a local *consent authority,* e.g. Tree Management Officer.

Court approved removal Removal of a tree subject to the granting of consent for such works under any law that provides for the removal of trees as ordered by a court of law.

Court determination Final decision made by a court of law after a *hearing* or trial.

Court order Direction/s given by a court of law.

Crack Narrow splitting along a stem, internal in origin, and may continue for some distance and depth (Mattheck and Breloer 1994, pp. 104–105). See also *Growth crack*.

Critically leaning A leaning tree where the trunk is growing at an angle greater than >45° from upright. See also *Leaning, Slightly leaning, Moderately leaning* and *Severely leaning*.

Slightly crooked   Moderately crooked   Severely crooked

Figure 7 Degrees of crookedness.

Critical roots See *Structural roots.*

Critical root zone (CRZ) A method that considers a minimum radial distance
from the trunk that disturbance to *structural roots* may occur for a tree to
remain stable. For this method, satisfactory setbacks are usually considered
as five times (5x) DBH with a minimum setback of 1.5 m for tree/s with a
trunk diameter of less than 300 mm (<300 mm DBH) with the possibility of
limited elevated construction as an incursion into the area after further
structural root examination, however, this does not consider *age, condition*
and *vigour.*

Crook See *Crooked.*

Crooked A stem with a bend or kink, with that atypical section orientated away
from its natural habit or *crown form*, usually returning to near its original
orientation further along the stem when light and space become available.
Bends may be a result of changes to original growing conditions or physical
damage and may impact on the tree's *stability* and structural integrity.
Degrees of crookedness can be categorised as *Slightly crooked, Moderately
crooked* and *Severely crooked.* (See Figure 7.)

Crooked tree See *Crooked.*

Crossed branches/roots Branches or roots forming a point of contact where they grow over each other. Such contacts may form a *graft* or a *root graft*, or may form a mechanical *abrasion wound* in branches subject to movement or a *false graft* in roots.

Crossover See *Vehicular crossover.*

Crotch The point on the inner side of a branch and trunk union or at the inner side of the union of two or more branches. The union may have an *acute branch crotch* forming a 'V' shape, an *obtuse branch crotch* or may be broadly rounded forming a 'U' shape. See also *Pocket crotch* and *Phytotelmata.*

Crotch of branch union See *Crotch.*

Crotch seam A *branch bark ridge* on the outer side/s of the crotch of a *compression fork* where *included bark* has become *enclosed bark* or partly enclosed evident as a narrow longitudinal protrusion along the branch's interface forming a weak *graft* (Mattheck & Breloer, 1994, pp. 64–65). See also *Occlusion seam* and *Rib.*

Crown Of an individual tree all the parts arising above the trunk where it terminates by its division forming branches, e.g. the branches, leaves, flowers and fruit; or the total amount of foliage supported by the branches. The crown of any tree can be divided vertically into three sections and can be categorised as *lower crown, mid crown* and *upper crown* (Figure 8). For a *leaning* tree these can be divided evenly into crown sections of one-third from the *base* to *apex*. The volume of a crown can be categorised as the *inner crown, outer crown* and *outer extremity of crown* (Figure 9).

Crown class See *Crown form.*

Crown cleaning See *Crown maintenance.*

Crown cover The estimated percentage of foliage covering the entire tree compared to that considered typical for the *taxon* when in *good condition* and of *normal vigour* and expressed as a percentage, considering *crown form* and *vigour, in situ.*

Crown density The estimated percentage of density of foliage present in the crown cover compared to that idealised for the genus and species when in

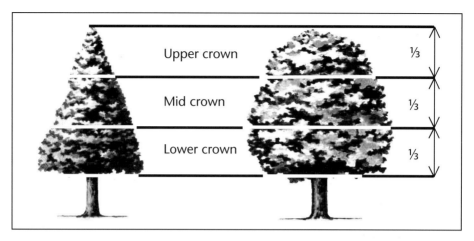

Figure 8 Crown sections.

good condition and of *normal vigour* and expressed as a percentage, considering vigour, predation by pests and diseases, environmental conditions such as drought, epicormic shoots and dormancy.

Crown form The shape of the crown of a tree as influenced by the availability or restriction of space and light, or other contributing factors within its growing environment. Crown form may be determined for tree shape and habit

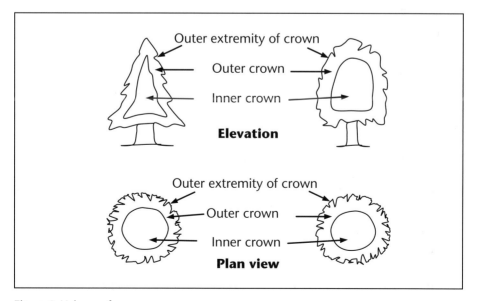

Figure 9 Volume of a crown.

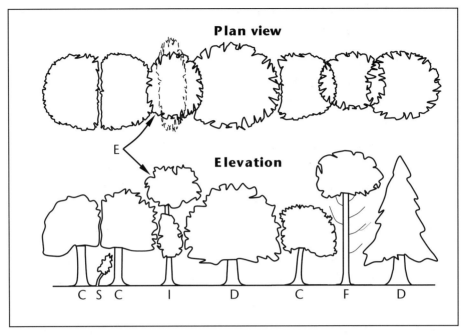

Figure 10 Crown form: D = Dominant; F = Forest; C = Codominant; E = Emergent; I = Intermediate; S = Suppressed. (Source: D, C, I and S, and Elevation, Matheny and Clark 1998, E, F and Plan View, IACA 2005)

generally as *dominant, codominant, intermediate, emergent, forest* and *suppressed* (Figure 10). The habit and shape of a *crown* may also be considered qualitatively and can be categorised as *good form* or *poor form*. See also *Forest grown* and *Open grown*.

Crown form codominant Crowns of trees restricted for space and light on one or more sides and receiving light primarily from above, e.g. constrained by another tree/s or a building.

Crown form dominant Crowns of trees generally not restricted for space and light receiving light from above and all sides. See also *Crown form emergent* and *Open grown*.

Crown form emergent Crowns of trees restricted for space on most sides receiving most light from above until the *upper crown* grows to protrude above the canopy in a stand or forest environment. Such trees may be *crown form dominant* or transitional from *crown form intermediate* to *crown form forest* asserting both *apical dominance* and *axillary dominance* once free of constraints for space and light.

Crown form forest Crowns of trees restricted for space and light except from above forming tall trees with narrow spreading crowns with foliage restricted generally to the top of the tree. The trunk is usually erect, straight and continuous, tapering gradually, crown often *excurrent*, with first order branches becoming structural, supporting the live crown concentrated towards the top of the tree, and below this point other first order branches arising radially with each *inferior* and usually temporary, divergent and ranging from horizontal to ascending, often with internodes exaggerated due to competition for space and light in the *lower crown*. See also *Forest grown*.

Crown form intermediate Crowns of trees restricted for space on most sides with light primarily from above and on some sides only. See also *Crown form emergent*.

Crown form suppressed Crowns of trees generally not restricted for space but restricted for light by being *overtopped* by other trees and occupying an understorey position in the canopy and growing slowly.

Crown integrity A percentage estimate of the extent of the crown that has died or is dying, an inverse estimate of *crown cover*.

Crown lifting Pruning to remove branches from the lower crown usually for clearance or access.

Crown maintenance Pruning that preserves the size and structure of a tree while maintaining crown volume (Australian Standard 2007, p. 6).

Crown modification Pruning that changes the structural appearance and habit of a tree (Australian Standard 2007, p. 6).

Crown projection (CP) Area within the *dripline* or beneath the lateral extent of the *crown* (Geiger 2004, p. 2). See also *Crown spread* and *Dripline*.

Crown projection area See *Crown projection*.

Crown protection zone (CPZ) A specified area above ground and at a given distance from the trunk, set aside for the protection of tree branches and foliage to provide for the viability and stability of a tree to be retained where it is potentially subject to disturbance by development. Note: Establishment of these areas may include pruning, tying-back of branches or other remedial works at the edge of the CPZ to prevent conflict between branches and works.

Crown raise See *Crown lifting.*

Crown reduction See *Reduction pruning.*

Crown regeneration An adaptation enabling the regrowth of the crown in some species to assert its natural habit and form after episodes of damage or severe stress, such as from fire, insects, lopping, storm damage, or drought. Such regrowth may be evident in a vigorous tree in the young to mature age stages of its life cycle.

Crownshaft A cylinder that crowns the top of a palm trunk that has formed from tightly packed tubular leaf bases which act as a protective measure for the *apical meristem* or growing apex, e.g. *Archontophoenix alexandrae.*

Crown shy An *asymmetrical* crown with exaggerated growth away from another tree or source of shading. See also *Crown form codominant.*

Crown spread The furthest expanse of the crown when measured horizontally from one side of the tree to the other, generally through the centre of the trunk. Where the crown is not circular a measurement should be an average of the narrowest and widest diameters, dependent on *crown form* and to a lesser extent its *symmetry.*

Crown spread orientation Direction of the *axis* of *crown spread* which can be categorised as *orientation radial* and *orientation non-radial.*

Crown spread orientation non-radial Where the crown extent is longer than it is wide, e.g. east/west or E/W. Further examples, north/south or N/S, and may be *crown form codominant*, e.g. A or B, *crown form intermediate*, e.g. A, or *crown form suppressed*, e.g. B*, and* crown symmetry is *symmetrical*, e.g. A, or *asymmetrical*, e.g. B. (See Figure 11.)

Crown spread orientation radial Where the *crown spread* is generally an even distance in all directions from the trunk and often where a tree has *crown form dominant* and is *symmetrical*. (See Figure 12.)

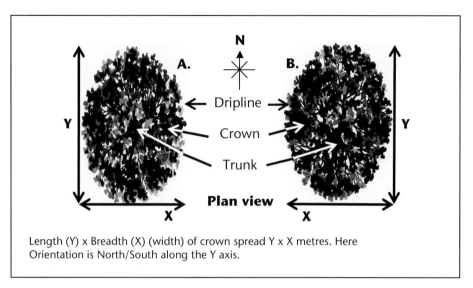

Figure 11 Crown spread orientation non-radial.

Crown thinning Removal of selected branches without modifying the size of a tree (Australian Standard 2007, p. 6).

Crown uplift See *Crown lifting.*

Cultivar (cv.) A contraction of *Cultivated variety.*

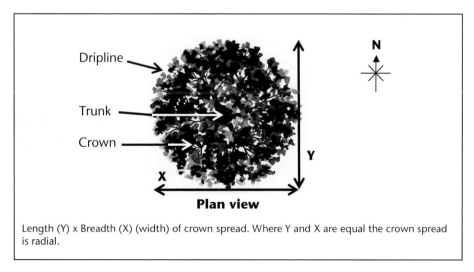

Figure 12 Crown spread orientation radial.

Cultivated variety (cv.) An assemblage of cultivated individuals (morphological, physiological, cytological, chemical etc.), significant for the purposes of agriculture, forestry and horticulture which when reproduced (sexually or asexually) retains its distinguishing features (Trehane *et al.* 1995).

Cultivation Any intervention to assist the growth of a tree or its progeny, usually to favour plants with specific characteristics.

Cupboard door branch failure (Mattheck & Breloer 1994, pp. 64–65) See *Collapse and hinge branch failure*.

Cup shakes See *Ring shake*.

Current season's growth *Growth increment* formed during an existing growing period. See also *Last season's growth* and *Growth rings*.

Curtain roots Free falling *aerial roots* descending to the ground usually from multiple points along a leaning host tree being strangled forming *column roots* supporting the trunk or branches, e.g. *Ficus virens* var. *sublanceolata*.

Curtilage The immediate area around a house or building enclosed by the land on which it is situated including out buildings, trees and landscape features.

Cut Removing of material to lower existing *grade*. See also *Fill*.

Cutting down See *Removal*.

D

Damping Dissipation of wind energy by a tree through resistance of the surface area of its leaves, branches and trunk by the swaying in complex looping movements of successively lower *orders of branches*, and movement of the reduced energy along the trunk to the *root plate* and into the soil (Lonsdale 1999, p. 313; James 2003, p. 167; and James 2005, p. 5).

Danger Potential for a tree's imminent *failure* and *collapse* in full or part, posing an *immediate risk* of *hazard* to the safety of people or damage to property. Danger is often a result of physical deterioration of a tree or tree part and its structure, or modification to the growing *environment* essential for its survival or physical *stability*.

Dangerous A tree or tree part that presents a *danger*.

Datum point Particular point from where all measurements or references are made. For a tree, it is considered that the *root crown* be the datum point as this position is relatively static throughout the life of the tree.

Dead Tree is no longer capable of performing any of the following processes or is exhibiting any of the following symptoms:

Processes
Photosynthesis via its foliage crown (as indicated by the presence of moist, green or other coloured leaves);

Osmosis (the ability of the root system to take up water);

Turgidity (the ability of the plant to sustain moisture pressure in its cells);

Epicormic shoots or *epicormic strands* in Eucalypts (the production of new shoots as a response to stress, generated from latent or adventitious buds or from a *lignotuber*);

Symptoms
Permanent leaf loss;

Permanent wilting (the loss of turgidity which is marked by desiccation of stems, leaves and roots);

Abscission of the *epidermis* (bark desiccates and peels off to the beginning of the sapwood).

Deadwood Dead branches within a tree's crown and considered quantitatively as separate to *crown cover* and can be categorised as *small deadwood* and *large deadwood* according to diameter, length and subsequent *risk* potential. The amount of dead branches on a tree can be categorised as *low volume deadwood, medium volume deadwood* and *high volume deadwood*. See also *Dieback*.

Deadwooding Removing of dead branches by *pruning*. Such pruning may assist in the prevention of the spread of *decay* from *dieback* or for reasons of safety near an identifiable target.

Decay Process of degradation of wood by micro-organisms (Australian Standard 2007, p. 6) and fungus.

Deciduous A woody plant, e.g. tree, shrub or vine, that sheds all of its leaves in one season and enters a *dormant* period, usually during winter.

Decline The response of the tree to a reduction of energy levels resulting from *stress*. Recovery from a decline is difficult and slow, and decline usually irreversible.

Decompaction Any process or procedure utilised to alleviate *soil compaction,* e.g. *radial trenching, vertical mulching,* or ripping prior to planting.

Decorticate The process where outer bark is shed from a tree annually as a result of calliper expansion by the addition of a new *growth increment,* e.g.

Corymbia citriodora (syn. *Eucalyptus citriodora*). Once fractured the old bark desiccates and shrinks causing further fragmentation and ultimately sloughs away. Note: A tree may be vulnerable to physical harm or to predation by pests and diseases before the new bark hardens off. See also *Persistent bark*.

Decurrent See *Deliquescent*.

Deep soil Soil to a depth of 1000 mm or more (Craul 1992, p. 32). See also *Shallow soil*.

Deep soil planting The concept of providing a sufficiently *deep soil* volume to support the growth of trees especially as part of landscape works for development. Note: this should also consider soil volume requirements for the lateral spread of the *root plate*.

Defect In relation to tree hazards, any feature of a tree which detracts from the uniform distribution of mechanical stress, or which makes the tree mechanically unsuited to its *environment* (Lonsdale 1999, p. 313).

Defence Any structure, system or process which defends tissues against damage (Lonsdale 1999, p. 313).

Deflecting Root barriers installed as wall-like or encircling solid obstacles to divert natural root direction, free an area of roots or direct roots to avert damage. Such barriers may be constructed of solid plastic, metal or timber (Coder 1998, p. 63; Roberts *et al.* 2006, p. 355).

Deflectors See *Root barrier* and *Deflecting*.

Deformation A change in form as a result of *loading stress* or from pests or pathogens in plant parts such as leaves and stems, e.g. galls.

Dehorning See *Lopping*.

Delaminate A *mechanical wound* caused when the bark is stripped from a tree, usually from the trunk as a continuous sheet back to the vascular cambium. This may occur from an impact or abrasion *episode* such as a collision with a motor vehicle and the tree may become *ringbarked*. See also *Partially delaminated*.

Delaminated joint Weakening of the union of branch and branch/trunk wood increments, where a dead branch has formed resulting in a weak union where each successive layers of branch/trunk *growth increment* no longer intertwine with the branch as the *branch tail* has ceased to grow. See also *Laminated joint*.

Delamination The separation of fibres often evident as longitudinal splitting of wood (Lonsdale 1999, p. 313).

Delegated authority Powers set in *legislation* for use by a *consent authority* given to staff members to undertake specific tasks.

Delignification The decomposition of *lignin* from *wood* by chemical deterioration, resulting in loss of strength, evident by separation of fibres into hairlike strands. See also *Lignification*.

Deliquescent Tree whose crown is comprised of two or more codominant *first order structural branches*. These branches can be categorised as *dual-leader*, *superior, inferior, permanent* or *temporary*.

Dendrochronology Study of time and past climates through the *growth rings* in timber. A separate system exists for the northern and southern hemispheres.

Dendrology The science of trees and study of woody plants (Wikipedia 2007).

Dendrophobia A recognised fear of trees or forests.

Densitomat® See *Penetrometer*.

Department Any administrative or operational section of government that is responsible for dealing with *policy* or administration, e.g. Environmental Planning.

Depth of margin Distance from outer trunk perpendicular to the *wound face*. This may assist in determining the age of a wound.

Descending hollow A *hollow* that develops downwards in a trunk or branch usually in a *proximal* direction. Such a hollow may be important for a *potential habitat tree*. See also *Hollow* and *Ascending hollow*.

Design level (DL) Level determined by the design process to which all data is referred.

Desire line An informal *footpath* formed as a worn track usually indicative of the shortest distance linking two regularly visited points in the landscape.

Destroy Any *immediate* or ongoing process or activity leading to the death of a tree.

Destructive root mapping See *Invasive root mapping.*

Detached broken branch/frond A live or dead branch or palm frond, that has snapped or fractured damaging its wood, destroying structural integrity at its point of connection, or from compartmentalisation by abscission (the frond), and has become separated from the tree at this point. Note: a detached broken branch or frond may pose a serious risk of hazard to the safety of people or damage to property, as its movement down through the crown may be unpredictable. (See Figure 2.)

Detritus See *Humus.*

Development The use of land, the subdivision of land, the erection of a building, the carrying out of a work, the demolition of a building or work and any other matter which is controlled by an environmental planning instrument.

Development application A formal submission to a *consent authority* by the property owner or their authorised agent describing the proposed development, and any impacts and their mitigation.

Development assessment The processes of evaluating a *development application* by the environmental and planning staff of a *consent authority.*

Development consent Formal notice of approval of *development application* with imposed conditions.

Development control plan (DCP) A plan prepared under the requirements of *legislation* and adopted by the consent authority, usually in addition to a planning document of broader significance, e.g. for issues of residential development, subdivision, parking, or specific places.

Diameter at breast height (DBH) Measurement of trunk width calculated at a given distance above ground from the base of the tree often measured at 1.4 m. The trunk of a tree is usually not a circle when viewed in cross-section, due to the presence of *reaction wood* or *adaptive wood*, therefore an average diameter is determined with a *diameter tape* or by recording the trunk along its narrowest and widest axes, adding the two dimensions together and dividing them by 2 to record an average and allowing the orientation of the longest axis of the trunk to also be recorded. Where a tree is growing on a lean the distance along the top of the trunk is measured to 1.4 m and the diameter then recorded from that point perpendicular to the edge of the trunk. Where a *leaning* trunk is *crooked* a vertical distance of 1.4 m is measured from the ground. Where a tree branches from a trunk that is less than 1.4 m above ground, the trunk diameter is recorded perpendicular to the length of the *trunk* from the point immediately below the base of the flange of the *branch collar* extending the furthest down the trunk, and the distance of this point above ground recorded as *trunk* length. Where a tree is located on sloping ground the DBH should be measured at halfway along the side of the tree to average out the angle of slope. Where a tree is *acaulescent* or *trunkless* branching at or near ground an average diameter is determined by recording the radial extent of the trunk at or near ground and noting where the measurement was recorded, e.g. at ground.

Diameter tape A tape for measuring trunk diameter where the markings are spaced at increments of 3.142 cm to correspond with ϖ (Pi) × distance (the formula for diameter) so that diameter does not need to be calculated (Farm Forest Line 2002).

Diametral cracks Longitudinal cracks formed on opposite sides of a stem with the potential of a *shear failure* (Lonsdale 1999, p. 49).

Diatropism A directional growth movement of a plant or plant part at right angles to a stimulus.

Dicot See *Dicotyledon*.

Dicotyledon The double embryonic seed leaf of some *angiosperms* that gives rise to flowering trees other than palms. See also *Cotyledon* and *Monocotyledon*.

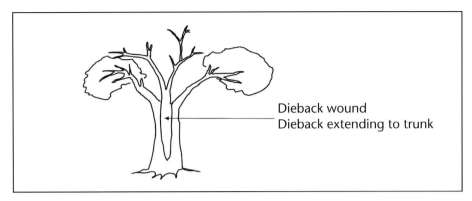

Dieback wound
Dieback extending to trunk

Figure 13 Dieback wound.

Dieback The death of some areas of the *crown*. Symptoms are leaf drop, bare twigs, dead branches and tree death, respectively. This can be caused by root damage, root disease, bacterial or fungal canker, severe bark damage, intensive grazing by insects, *abrupt changes* in growth conditions, drought, water-logging or over-maturity. Dieback often implies reduced *resistance, stress* or *decline* which may be temporary. Dieback can be categorised as *Low volume dieback, Medium volume dieback* and *High volume dieback.*

Dieback wound Wounding where *dieback* extends beyond a branch collar as with *natural pruning* and extends to other branches, trunk or roots. See also *Secondary crown* and *Stag-headed*. (See Figure 13.)

Diffuse bundles *Vascular cambium* sporadically distributed in palm stems and separated by *ground tissue*. Palm stems are not subject to *secondary thickening* and therefore do not form *growth rings* as with *gymnosperms* and dicotyledonous *angiosperms.*

Diffuse porous One of the two main types of wood structure in broadleaf trees in which the diameter of the vessels decreases progressively from the *early season's wood* to the *late season's wood,* e.g. Cherry, Aspen, Poplar. See also *Ring porous.*

Directed growth The cultivated or guided development of root growth in a specific location or direction, usually to establish new roots or protect infrastructure, e.g. *baiting* or *channelling* (Coder 1998, p. 65).

Directional pruning See *Selective pruning.*

Discoloured wood Injury altered wood chemically changed initially by the tree for protection modifying its colour and then from a succession of interactions between the tree with infesting micro-organisms and then between competing micro-organisms (Shigo 1986, p. 35).

Disease A malfunction in or destruction of tissues within a living organism, usually caused by pathogenic micro-organisms (Lonsdale 1999, p. 313) and environmental factors.

Distal A section of any tree part, furthest from its point of attachment. See also *Proximal.*

Divergent branch The direction taken by a branch as it grows away from another. See also *Convergent branch.*

Dominance A tendency in a leading shoot to maintain a faster rate of *apical* elongation and expansion than other nearby lateral shoots, and the tendency also for a tree to maintain a taller crown than its neighbours (Lonsdale 1999, p. 313).

Dominant epicormic See *Elite.*

Dormancy See *Dormant.*

Dormant A state of reduced cellular activity (Capon 1990, p. 209).

Dormant bud An inactive bud that forms in the axils of leaves and remains in the *cambial zone* as the tree grows (Shigo 1989c, p. 134).

Dormant tree vigour Determined by existing turgidity in lowest order branches in the outer extremity of the crown, with good bud set and formation, and where the last extension growth is distinct from those most recently preceding it, evident by bud scale scars. Normal vigour during dormancy is achieved when such growth is evident on a majority of branches throughout the crown.

Drainage cells A proprietary system of durable plastic construction providing drainage and structural support to relieve *hydrostatic pressure* behind walls and drainage under paving.

Drainage plan See *Hydrology plan.*

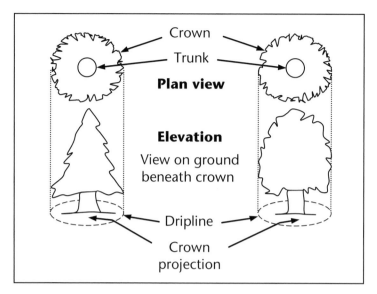

Figure 14 Dripline and crown projection.

Drey Nests in trees made by possums or squirrels.

Dripline A line formed around the edge of a tree by the lateral extent of the *crown* (Figure 14). Such a line may be evident on the ground with some trees when exposed soil is displaced by rain shed from the crown. See also *Crown projection*.

Driveway Area set aside for vehicular access from a roadway to a property, usually across a road reserve.

Dropcrotch cuts Pruning usually in *young* trees to reduce *height* and crown spread and slow growth by reducing stem length to *laterals* of proportionate diameter, usually one-half ($^1/_2$) to rarely one-third ($^1/_3$) that of the stem being removed (Gilman 1997, pp. 22–25).

Dropper roots See *Sinker roots*.

Drop zone The distance away from a tree that may be physically influenced by a falling branch.

Drought A prolonged period of little or no rain.

Drought stress See *Permanent wilting point*.

Dry face See *Wound face.*

Dual leader branches A *deliquescent* crown comprised of two *Codominant first order structural branches* (FOSB) where each branch has a similar diameter or supports an approximately even percentage and volume of *crown cover.*

Dual leaders See *Dual leader branches.*

Dwarf mistletoe See *Mistletoe.*

Dynamic load Loading force that is moving and changes over time, e.g. from wind movement (James 2003, p. 166). See also *Static load.*

Dysfunctional xylem *Xylem* that is inactive, diminished or failed, unable to conduct materials.

Dysfunction The loss of physiological function, especially water conduction (Lonsdale 1999, p. 313).

Early season's wood In some trees, large cells with numerous vessels but thin walls produced as a consequence of rapid growth as day length and temperature increases during spring and summer to pump the maximum amount of water for growth. See also *Late season's wood*.

Early wood See *Early season's wood*.

Easement Areas of land above or below ground, subject to statutory constraints being either public or private provided for access or the location of utilities.

Eccentric cavity See *Asymmetrical cavity*.

Ecological community A group of organisms occurring in a particular area.

Edge tree 1. A tree located at the outer extremity of a stand either naturally occurring or planted and exposed to extra sunlight and wind on at least one of its aspects, usually resulting in *atypical* growth when compared to other trees within the stand causing a growth habit and *crown form* to develop that adjusts to the conditions. Such trees usually form a protective perimeter to the stand filtering and slowing the speed of wind to the inner trees. 2. A tree exposed by clearing or other modification to a stand, e.g. storm *episode*, now placing it at the outer extremity of the stand and vulnerable to exposure and *windthrow* as a result of it having initially developed a growth habit or *crown* *form* different from that of a tree naturally occurring or planted at the outer extremity of a stand.

Edge of trunk (EOT) A specified point on the trunk from where a measurement may be taken, e.g. the boundary was 2.5 m from EOT, to south. See also *Centre of trunk.*

Electrical conductivity meter An electronic device that measures the electrical resistance of wood utilising *ion* concentrations from water content (Nicolotti & Miglietta 1998, pp. 297–302). Examples of such proprietary devices are the Shigometer® and Vitamat®.

Elephant ears rib See *Round-edged rib.*

Elevating work platform (EWP) Any ground based mechanical device providing access to the *crown* of a tree for *assessment* or *maintenance,* e.g. travel tower or crane.

Elevation plan A plan viewing a design from a particular side, e.g. north elevation.

Elite An *epicormic shoot/s* that grows to become dominant from a *sprout mass.* See also *Inferior epicormic.*

Embedded bark See *Included bark.*

Embrittlement See *Structural deterioration.*

Emergency removal Tree *removal* as a matter of urgency due to its imminent potential to cause damage to people or property as a result of structural defects or modifications to its growing *environment* rendering it vulnerable to failure in full or part, e.g. a tree in adverse weather conditions suddenly developing a *progressive lean* and collapsing across a busy road.

Enclosed bark 1. Bark concealed within the wood of a *compression fork* after the grafting or intertwining of branches above the crotch in an *acute branch union*. 2. Bark concealed by *occluded* wounds, or where *convergent branches* grow into contact above the original crotch of a branch union where persistent bark has accumulated. On branches this may be more prevalent where the angle in the crotch of the branch union is acute i.e. less than 90° and more commonly on ascending and upright branches.

Enclosed wound Wound with a perimeter of *wound wood* with a well-defined apex, base and margins and often evident on an older wound. On a pruned branch that is rounded the enclosing wound wood from the branch collar may be circular with no definite apex or base evident. However, on a pruned branch where the *wound face* is oval in shape due to *reaction wood*, the enclosing *wound wood* from the branch collar may form a definite apex, base or margins.

Encroachment The growth of branches, trunk or roots onto another property.

Endemic A *native* plant usually with a restricted occurrence limited to a particular country, geographic region or area and often further confined to a specific *habitat*. See also *Indigenous*, *Locally indigenous* and *Non-locally indigenous*.

End loading The result of incorrect pruning practice where foliage is concentrated at the end of first order structural branches by removal of lower order branches, and may cause detrimental stress loading problems by removing the *damping* protection provided by the lower order branches (Lonsdale 1999, p. 37).

Endodermis The layer of cell/s at the innermost edge of the cortex forming a boundary with the stele in younger dicots before the vascular bundles merge to form the cambial ring, providing a waterproof band of cell/s that harden with age providing an outer layer of protection for the vascular cambium.

End weight Excessive formation of foliage concentrated at the *distal* end of a branch.

Entwined branches Two or more branches growing so that one or more twists around the other, with or without forming a *graft*.

Environment The conditions in which a tree lives. This includes the *biotic* and *abiotic* factors such as the effects of other living organisms and the non-living physical and chemical factors such as light, water, nutrients, soil and weather.

Environmental impact assessment See *Environmental impact statement*.

Environmental impact statement (EIS) A study to identify specific environmental impacts of a development proposal usually submitted with a development application.

Environmental law Legislation underpinning a land use planning system, e.g. in New South Wales, Australia, the *Environmental Planning and Assessment Act 1979.*

Environmental planning instrument Any policy or plan governing environmental issues, e.g. in New South Wales, Australia, State Environmental Planning Policy (SEPP), Regional Environmental Plan (REP) and Local Environmental Plan (LEP).

Environmental wounding/damage Wounding inflicted by environmental factors or modifications to the growing *environment* of a tree, e.g. sun-scald, drought, fire, water logging, wind damage to leaves, branches, bark or roots, phytotoxic damage from chemicals, or air, soil or water pollution.

Ephemeral A plant or plant part living for only a short period of time.

Epicormic Shoots arising from *latent buds* or *adventitious buds*. See also *Epicormic shoots* and *Epicormic buds*.

Epicormic bud Buds produced from *epicormic strands* in Eucalypts (Burrows 2002, pp. 111–131), or buds, shoots or flowers borne on old wood, usually applied to shoots arising after injury, e.g. after fire on eucalypts, from dormant buds.

Epicormic meristem strands See *Epicormic strands*.

Epicormic shoots Juvenile shoots produced at branches or trunk from *epicormic strands* in some Eucalypts (Burrows 2002, pp. 111–131) or sprouts produced from dormant or latent buds concealed beneath the bark in some trees. Production can be triggered by fire, pruning, wounding, or root damage but may also be as a result of *stress* or *decline*. Epicormic shoots can be categorised as *low volume epicormic shoots*, *medium volume epicormic shoots* and *high volume epicormic shoots*.

Epicormic stem Branch derived from an *epicormic shoot*.

Epicormic strands In some taxa of the *Myrtaceae* family narrow bands of *meristematic* tissue radiate in stems from *pith* extending to the outer bark containing bud primordia evident as small prickle or dimple like structures up to 10 mm diameter, that after the stimulus of a trauma *episode* such as fire or defoliation develop to form new buds allowing *crown regeneration* (Burrows 2002, pp. 111–131).

Epidermis In bark the outer layer of living cells on a stem that protects the underlying tissue from moisture loss and attack by pathogens.

Epiphyte A plant that utilises another as a host for growth and support without adversely impacting upon that host, depending largely on nutrients from humus formed from leaf litter accumulated in a crotch or hollow, or nutrients washed down along branches or trunk of the host, e.g. some ferns and orchids.

Epiphytic See *Epiphyte.*

Episode A single occurrence of importance that may be repeated, or one of a series of incidents or events that may impact upon a tree or its growing environment, e.g. drought followed by fire then *predation* by insects or disease; soil level changes (e.g. filling/excavation) and root severance followed by excessive pruning or lopping.

Erect A branch growing in an upright direction away from its point of attachment.

Escape A plant introduced into cultivation that produces viable *progeny* and spreads into other areas and is able to become established and may inhibit the growth of existing vegetation or restrict or suppress the viable progeny of *remnant indigenous* vegetation or cultivated plants.

Espalier Pruning and training of a tree to grow flat against a wall or on a trellis (Australian Standard 2007, p. 6).

Essential elements Soil elements in solution derived from mineral solids, air and water required for the growth of plants. From air and water Carbon, Hydrogen and Oxygen are required in the largest amounts and from the soil, minerals in the form of *macronutrients* and *micronutrients.*

Evaporation The return of moisture to the atmosphere in a gaseous form, from soil or a body of water.

Evapotranspiration A contraction of *evaporation* and *transpiration* recognising the collective effects of those processes.

Event See *Episode*.

Evergreen A tree that retains its leaves throughout the year.

Evidence Documents or other material presented as proof to support an argument to be determined in a court of law.

Excluded bark See *Extruded bark*.

Exclusion zone A *soil profile* physically or chemically altered to prevent the growth of roots, e.g. by *soil compaction* or chemical treatments (Coder 1998, p. 62).

Excrescence Outgrowths or enlargements on a tree, usually abnormal, e.g. *burl*, *gall* (Boland *et al.* 2006, p. 688).

Excurrent Tree where the trunk is erect, straight and continuous, tapering gradually, with the main *axis* clear from base to apex, e.g. *Araucaria heterophylla* – Norfolk Island Pine. Note: some tree species of *typical* excurrent habit may be altered to *deliquescent* by physical damage of the *apical meristem*, or from top lopping, or from the propagation of inferior quality stock. However, *formative pruning* may be able to correct a *crown* to excurrent if undertaken when a tree is *young*.

Exempt development Minor forms of development exempt from approval by the *consent authority* providing they satisfy certain planning criteria.

Exerted bark See *Extruded bark*.

Ex-ground trees Tree/s specifically grown in the ground then later excavated for use as part of advanced landscaping stock. Such trees may include *balled-in-burlap* and *in-ground container* (Clark 2003, p. 10).

Exit hole A hole in the trunk, roots or branches as a result of borer insects living inside the tree, reaching maturity and exiting by boring to the outside of the tree usually to reproduce. Holes vary in shape and size.

Exotic A plant introduced from another country or region to a place where it was not *indigenous*. Such plants may become *naturalised* and often originate as garden *escapes*.

Exotropy The ability of a root deflected by an obstacle to resume the orientation of its original direction once it has grown past the obstruction.

Expert witness conferencing Meeting/s or communications between expert witnesses with the consent of a court of law to determine which matters of dispute in a case can be resolved prior to the *hearing* to expedite matters allowing the court to deal with those matters remaining in dispute and submitting a summary of the outcomes of communications as a *statement of agreed facts.*

Expert testimony The *evidence* provided by an *expert witness* given under oath or deposition in a court of law to assist the court with the making of its determination in the case.

Expert witness A knowledgeable and experienced individual in a particular field who provides *evidence* or assistance to a court with the making of its determination in the case.

Extant Still existing, e.g. soil levels, *remnant vegetation.*

Extension growth Increase in length of a stem or root in one growing period, allowing for several growing periods that may occur annually on some trees. Note: Variation in lengths on a stem may be beneficial as a diagnostic tool. See also *Growth increment.*

Extruded bark The bark in the crotch of a branch union or *wound* that is incrementally shed from the tree by each successive *growth increment.* On branches the extruded bark may be evident as a protrusion or striation of the bark forming a ridge as the branch collar traverses the crotch of the branch union.

Exudate Oozing of *sap* from severed or ruptured *vascular cambium.*

Fall zone The distance away from a tree that may be physically influenced if it was cut down or subject to *collapse*.

Fail See *Failure*.

Failure The structural collapse in part or full of a branch or tree that has been physically diminished by wounding or from the actions of pests and diseases, or overcome by loading forces in excess of its load-bearing capacity including the subsequent loss of soil cohesion, respectively.

Fair condition Tree is of good habit or *misshapen*, a form not severely restricted for space and light, has some physical indication of *decline* due to the early effects of *predation* by pests and diseases, fungal, bacterial, or insect infestation, or has suffered physical injury to itself that may be contributing to instability or structural weaknesses, or is faltering due to the modification of the *environment* essential for its basic survival. Such a tree may recover with remedial works where appropriate, or without intervention may stabilise or improve over time, or in response to the implementation of beneficial changes to its local environment. This may be independent from, or contributed to by vigour. See also *Condition*, *Good condition* and *Poor condition*.

False graft Contact between two or more roots where a *graft* appears to have formed, but has not.

Fasciation Abnormal growth where a shoot becomes enlarged and flattened appearing as fused shoots possibly caused by virus or mycoplasma.

Feasance To carry out an official duty, or legal obligation as required. See also *Nonfeasance*.

Feathering See *Lion's tailing*.

Feature planting See *Specimen tree*.

Feature tree See *Specimen tree*.

Feeder roots An inappropriate term used to describe *fine roots*. It is inappropriate because plants are *autotrophs* being organisms that produce their own food through photosynthesis, and therefore do not feed but instead produce their own food.

Fiber See *Fibre*.

Fibre A type of vertically aligned axially elongated wood cell with a narrow lumen and thick robust wall, providing strong mechanical support; also, a general term for the axially elongated cells in wood (Lonsdale 1999, p. 314).

Fibrous roots See *Fine roots*.

Field capacity After *field saturation* this is the remaining water in the ground that cannot be freely drained by gravity.

Field-grown See *Ex-ground trees*.

Field saturation Maximum amount of water able to be stored in soil when all air pores are replaced with water.

Fill Material such as soil, rock and builder's waste introduced to a location to raise the height of levels, e.g. to above existing *grade* or to deposit sufficient material to a natural or artificial depression to restore it to a level consistent with or above the surrounding landscape.

Final cut The last pruning cut in the process of the reduction or removal of branches. The purpose of this cut is to reduce the risk of micro-organism infection according to the principles of *compartmentalisation* and to encourage wound *occlusion* (Australian Standard 2007, p. 7).

Fine frass *Frass*, powder like to the touch with particles less than <0.1 mm in diameter. See also *Frass*, *Coarse frass* and *Medium frass*.

Fine roots 1. Lowest order of non-woody roots usually <1–2 mm long and 0.2–1 mm or less in diameter, responsible for absorption of water and nutrients in solution. Elongation occurs at the *root tip* of these roots and such roots may be short lived or persist (Perry 1982, pp. 197–221). Fine Roots are often erroneously referred to as *feeder roots*. 2. Roots that may arise where the *radicle* is replaced by *adventitious roots* branching many times as with palms or grasses or other monocotyledons.

Fire wound Wounding caused by fire. Such wounds may cause initial damage or may be secondary from a previous wounding *episode*/s. Some fire damage may be superficial or may destroy a tree in full or part rendering it potentially vulnerable to failure. Note: fire damaged trees can be potentially hazardous and should be assessed carefully.

First order branch (FOB) Initial branch arising from the *trunk* or *root crown*. Such a branch may be *structural* or *non-structural*, *temporary* or *permanent*, as a *codominant dual-leader branch*, *superior* or *inferior*, forming a *crown* of *deliquescent* habit.

First order roots (FOR) Initial woody roots arising from the *root crown* at the base of the *trunk*, or as an *adventitious root mass* for structural support and *stability*. Woody roots may be buttressed and divided as a marked gradation, gradually tapering and continuous or tapering rapidly at a short distance from the root crown. Depending on soil type these roots may descend initially and not be evident at the root crown, or become buried by changes in soil levels. Trees may develop 4–11 (Perry 1982, pp. 197–221) or more first order roots which may radiate from the trunk with a relatively even distribution, or be prominent on a particular aspect, dependant upon physical characteristics, e.g. leaning trunk, *asymmetrical* crown; and constraints within the growing *environment* from topography, e.g. slope, soil depth,

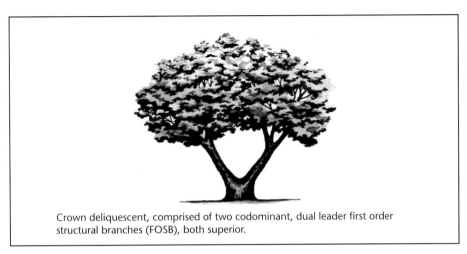

Crown deliquescent, comprised of two codominant, dual leader first order structural branches (FOSB), both superior.

Figure 15 Codominant, dual leader first order structural branches (FOSB).

rocky outcrops, exposure to predominant wind, soil moisture, depth of *water table* etc.

First order structural branches (FOSB) A branch or branches arising from the trunk to form the initial *orders of branches* elongated to develop a permanent framework of branches supporting the *crown*, usually persisting beyond the tree's maturity (Figures 15–18.)

Fissure A seam between concave edges of *fluted* sections of a trunk, branch or root.

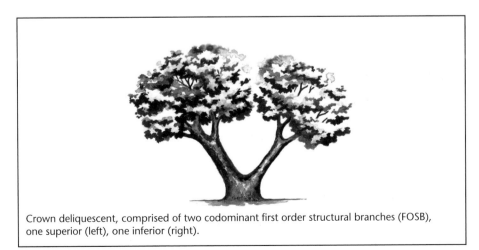

Crown deliquescent, comprised of two codominant first order structural branches (FOSB), one superior (left), one inferior (right).

Figure 16 Codominant first order structural branches (FOSB).

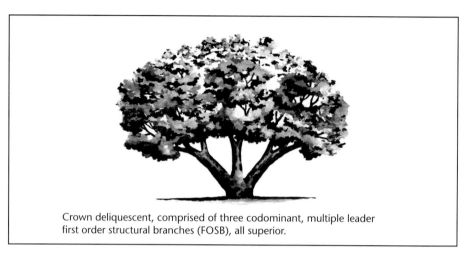

Crown deliquescent, comprised of three codominant, multiple leader
first order structural branches (FOSB), all superior.

Figure 17 Codominant, multiple leader first order structural branches (FOSB).

Flagging 1. *Asymmetrical crown form* as a result of exposure to dominant pre-vailing winds where the crown is modified by its growing conditions until it is aligned like a flag on a pole away from the wind and along its flow path (Ennos 2001, p. 44), e.g. trees growing in exposed locations such as coast-lines and cliffs. 2. Death of separate *lower order branches* in the *outer extremity of crown*, such damage may be caused by Cockatoos (Parrots). See also *Krummholtz form*.

Flare See *Basal flare*.

Crown deliquescent, comprised of three codominant first order structural branches (FOSB),
two superior (left and right), one inferior (centre).

Figure 18 Codominant, dual leader, multiple first order structural branches (FOSB).

Flashback See *Backflash.*

Flash-back See *Backflash.*

Flatroot See *Shallow rooters.*

Flower bud A bud that produces a flower.

Flowering plants Applied generally to *angiosperms.*

Flush cut An incorrect cut that damages or removes the *branch collar* or *branch bark ridge* and as a result damages stem tissue.

Fluted A section of trunk, branch or root that is broadly convex or cable-like and may be linear, *helical* or interconnected with sections usually separated by a *fissure.*

Foliage crown See *Crown.*

Footing Of a building, the lowest part of a structural wall or *pier* (in any form), that rests upon or into the earth. See also *Foundation.*

Footpath A dedicated pedestrian access way, usually constructed of concrete.

Force Influence that causes something to move.

Forest grown A tree with *crown form forest* grown in a group with competition for space and light protected from wind, often resulting in a taller tree with a narrow spreading crown that is concentrated towards the top of the tree (Matheny & Clark 1998, p. 18). See also *Open grown,* and Figure 20 on p. 104.

Forestry See *Silviculture.*

Fork *Bifurcation* of a branch or root into lower *orders of branches* or *orders of roots.*

Form See *Crown form.*

Formative pruning The pruning of young trees usually, to assist with the development of *crown form* and shape and to develop strong structure (Australian Standard 2007, p. 7). Such pruning may reduce developmental weaknesses, e.g. *crossed branches,* branches with branch bark inclusions, or to remove codominant first order branches to extend the length of a trunk or to guide the *crown form* of a tree to a single first order structural branch, or to encourage branching to make a crown shape excurrent or deliquescent, respectively.

Forty-five degrees (45°) crossed ribs *Wound wood* having *occluded* 45 degree slip lines (Mattheck 2004, p. 59).

Forty-five degrees (45°) slip lines Intersecting lines of *shear failure* in the 45 degree plane as a result of *compressive loading* in vertical stems generally observed at old branch knots. Failure may result along a line of slippage across the vertical stem like an angle of repose, or by the intersecting 45 degree lines forming a downward wedge opening the stem along its length (Mattheck 2004, pp. 52, 58–59).

Foundation The point at the earth surface upon which a *footing* rests. See also *Footing*.

Fractometer® Proprietary device to measure the breaking strength and flexibility of *increment core* samples extracted with a *Pressler increment borer* (Mattheck & Breloer 1994, p. 191).

Framework roots See *Structural roots*.

Frangible Tree and shrub *taxa* utilised as roadside plantings and favoured as a result of their predisposition to break upon impact, especially from motor vehicle accidents. Such trees are usually of small dimensions or often *shrub* like.

Frass The granular wood particles produced from borer insects and can be categorised as *Fine frass*, *Medium frass* and *Coarse frass* with the different types being of different sizes and caused by different insects.

Frizzle top Consequence of manganese deficiency causing chlorotic, withered or diminished new leaves on palms, often as a result of planting too deep (Shigo 1991; Harris *et al.* 2004).

Fruit tree A tree grown as an *ornamental* or commercially for the production of edible fruit.

Functional wood Living wood, e.g. sapwood.

Fungal spores The reproductive structures of fungi. Spores differ in shape and colour, are microscopic and are one or many celled (Manion 1991, p. 109).

Fungi Plural of *Fungus*.

Fungus In trees a *heterotrophic* organism that grows by decomposing cellulose and *lignin* from a living host often to the detriment of the host – parasitic fungus, or decomposes dead organic matter – saprophytic fungus, or grows in a symbiotic association with roots. Fungus detrimental to its host is a pathogen and can cause diseases such as wood *decay*, rust diseases, *canker* diseases, foliage diseases, wilt diseases and *root rot* diseases. A fungus grows as branching and filamentous thread-like structures called *hyphae* masses of which form the vegetative body of the fungus called mycelium. The tips of the hyphae grow between cells and within cell walls and permeate the cell releasing digestive enzymes that make the invaded material soluble and this is then diffused through the cell wall of the hyphae and metabolised through the *mycelium* (Manion 1991, pp. 102–109). There are two main groups of fungi, Microfungi which is only visible with a microscope and Macrofungi which can be seen by the unaided eye. The Macrofungi are placed into three divisions called the *Ascomycota*, *Basidiomycota* and *Myxomycota* (slime moulds) (Young & Smith 2005, pp. 1, 2, 5–7).

Gall Abnormal localised swelling or an outgrowth on a leaf, stem or root, caused by a parasite.

Gap graded fill A rock fill of a given aggregate size, e.g. 40 mm diameter gravel with no fines, as a media added above existing *grade* when soil levels are to be increased near existing trees, to provide sufficient *pore* spaces between fill particles allowing for *gaseous exchange* between tree roots, soil and the atmosphere. This may be a layer upon which silt fabric is spread and other fill media or surface treatment such as a load-bearing concrete slab placed on top with the capacity usually determined by the type of material for strength and durability and aggregate sizes.

Garden escape See *Escape*.

Gaseous exchange The transfer of gases between the tree and its growing environment.

Geocast wall A proprietary system for the construction of concrete walls cast in the ground primarily to *shore* basement excavations in limited space reducing the need for *over excavation* and *benching* usually required with more conventional excavation and *shoring* methods.

Geotechnical engineer An expert in evaluating foundations for buildings, roads and other structures, and their interaction with soil and geology.

Geotropism See *Gravitropism*.

Girdle scar See *Bud scale scar*.

Girdling root The growth of a root that is not radial away from the trunk or root/s and curves to encircle the trunk or root/s constricting phloem or *vascular cambium* causing dysfunction. This is usually caused by the roots of plants being confined in circular growing containers for too long, where the radiating roots reach the edge of the pot and are deflected around its curved surface, or from seedlings being potted on incorrectly, e.g. forming a 'J' curved root. This is also occasionally caused when larger established roots deflect the growth of a new or smaller root away from its radial predisposition, and poor preparation of a planting hole in a hostile growing environment. See also *Exotropy*.

Good condition Tree is of good habit, with *crown form* not severely restricted for space and light, physically free from the adverse effects of *predation* by pests and diseases, obvious instability or structural weaknesses, fungal, bacterial or insect infestation and is expected to continue to live in much the same condition as at the time of inspection provided conditions around it for its basic survival do not alter greatly. This may be independent from, or contributed to by vigour. See also *Condition*, *Fair condition* and *Poor condition*.

Good form Tree of *typical* crown shape and habit with proportions representative of the taxa considering constraints such as origin, e.g. indigenous or exotic, but does not appear to have been adversely influenced in its development by environmental factors *in situ* such as *soil water* availability, prevailing wind, or cultural practices such as lopping and competition for space and light. See also *Poor form*.

Good vigour See *Normal vigour*.

Grade Ground level at a specified point.

Graft The permanent connection of vascular tissue usually between different specimens of the same species or parts of the same tree, either incidentally or deliberately for *propagation* purposes, where their live roots or stems grow against one another with sufficient pressure to enable living cells to inter-twine allowing the movement inter-plant of plant growth regulators (hor-mones) through the *phloem* (Thomas 2000, pp. 94 and 97) and assimilates.

Grafted branches Where two or more convergent or crossed or *entwined branches* from the same tree or another tree of a related species grow and merge forming a permanent union sharing vascular functions and structural loading. This may be deliberate as a horticultural process, e.g. *pleaching*, or incidental, e.g. *Angophora costata, Lagerstroemia indica*.

Grafted roots See *Root graft*.

Grafted root zone Multiple *root grafts* in entangled and overlapping *root plates* of trees of the same species and sometimes genus growing proximate, or in a forest environment.

Grafted stems See *Grafted branches*.

Grain The direction, size and arrangement of *fibre* in wood.

Gravitropism A tropism exhibited in response to gravity, resulting in stems growing upwards (negative tropism) and roots growing downwards (positive tropism).

Ground penetrating radar *Radar* transmitted into the ground from an antenna reflected to the surface off some objects to a receiving antenna where the strength and time of the returning signal is processed by computer enabling the formation of images of below ground structures such as tree roots, cables and pipes.

Ground tissue In palm stems the region between *diffuse bundles* of *vascular cambium* comprised of thin walled *parenchyma* cells (Shigo 1989b, p. 29) as *pith* for the storage of nutrients.

Ground water See *Underground water*.

Ground water level The upper limit of a body of *underground water* after pores in the underlying rock strata are filled.

Growing media A variety of composite materials for growing plants in the ground or in containers that may be derived from a combination of natural *soil*, *organic* or *inorganic* material, providing mineral particles, organic matter, water, air and living organisms to continuously supply roots with balanced proportions of water, air and nutrient elements (Handreck & Black 2002, p. 6).

Growing medium Singular of *Growing media*.

Growth crack Longitudinal split that may develop as a rupture in the bark from normal growth. See also *Growth split*.

Growth increment Wood layer formed during a growing period. See also *Growth rings*.

Growth rings In some trees *secondary thickening* forms distinct concentric bands evident in the wood produced on the outer side of a stem during growing periods by alternating layers of *Late season's wood* and *Early season's wood*. Where trees are deciduous and lie dormant over winter the changes between the bands may be stark or slight or almost not apparent for some evergreen and rainforest trees.

Growth split Longitudinal crack that may develop in the trunk of some fast growing palms (Jones 1996, p. 268).

Gum See *Kino*.

Gutter The usually concrete or stone, flattened edge adjacent to the base of a kerb extending for a short distance into and usually level with the edge of the roadway, as a surface drain of the roadway and from the storm water pipes from adjoining properties.

Guying A form of artificial support using *bracing* for trees where their anchorage is temporarily inadequate (Lonsdale 1999, p. 315).

Gymnosperms Plants where the ovule (seed) is not enclosed fully within the fruit, i.e. the seeds are naked. These trees form cone flowers, often have woody *cone* fruit, are known as *conifers* and are generally referred to as softwood trees although some have hard durable wood. See also *Angiosperms*.

Habit The shape of a tree when its growth is unencumbered by constraints for space and light, e.g. idealised by an isolated *field grown* specimen with consideration of the species and the type of *environment* in which it evolved.

Habitat An area occupied, or periodically or intermittently occupied, by a species, *population* or *ecological community* including any *biotic* or *abiotic* component.

Habitat tree Any tree providing a niche supporting the life processes of a plant or animal, e.g. a *hollow* in the trunk or branches, suitable for nesting birds, arboreal mammals and marsupials, e.g. squirrels, bats or possums, or support of the growth of epiphytic plants, e.g. orchids, ferns. See also *Potential habitat tree*.

Hair roots See *Fine roots*. See also *Root hairs*.

Halophyte A plant adapted to living in salty conditions.

Hand excavation Digging undertaken with non-motorised hand tools near an existing tree, usually to locate and protect roots.

Hanger See *Hanging branch*.

Hanging branch *Detached broken branch* remaining within the *crown* by being tangled or supported by branches within the crown, or the crown of a nearby tree or built structure.

Hardwood See *Angiosperms.*

Hat-racking See *Lopping.*

Haustoria Plural of *Haustorium.*

Haustorium In *mistletoe* the collective term for sinker cells once established in the xylem, and cortical strands of cells once established in the *cortex* and phloem of the stem of the host to extract water and nutrients (Coder 2004, pp. 37–44).

Hazard The threat of *danger* to people or property from a tree or tree part resulting from changes in the physical condition, growing environment, or existing physical attributes of the tree, e.g. included bark, soil erosion, or thorns or poisonous parts, respectively.

Hazard abatement Action taken to reduce the potential failure of a tree in full or part or disruption by its growth to built structures, decreasing the *risk* of injury to people or damage to property.

Hazard assessment A tree assessment to determine the structural integrity, *stability*, *viability* or suitability of a tree for its retention *in situ*, remedial works or removal by identifying and analysing potential targets and the likely risk for failure or collapse in full or part, or disruption to growth, affecting those targets over *periods of time.*

Hazard beam Occurs where a stem that curves upwards is bent in the opposite direction to the curve as a result of excessive loading forces causing a longitudinal split along the stem (Mattheck 1999, p. 30).

Heading cut Cutting of a branch between the nodes (Gilman 1997, p. 25). See also *Lopping.*

Healing Physiological processes not known to occur in trees (Gilman 1997, p. 170). See also *Occlusion* and *Wound.*

Health A tree's *vigour* as exhibited by *crown density*, *crown cover*, leaf colour, presence of epicormic shoots, ability to withstand *predation* by pests and diseases, *resistance* and the degree of *dieback*.

Hearing A time and date/s set aside by a court of law for parties in dispute to present evidence, allow legal counsel representing the interests of opposing parties and the judge or *commissioner* to question witnesses under oath and assist the court with the making of its determination in the subject case.

Heart rooters Trees with a *root plate* morphology where *heart roots* develop for strong *anchorage* and buttresses for *stability*. Such trees may be subject to *windthrow* where their reliance on soil friction is overcome by prolonged rain periods reducing critical *shear stress* between the soil and roots (Mattheck & Breloer 1994, pp. 70, 74).

Heart roots Descending *structural roots* growing at an angle from the *root crown* or *buttress* other than along the soil surface or vertically, and expanding to provide *anchorage* and *stability* to the tree (Harris *et al.* 2004, p. 529).

Heart rot *Decay* occurs in the centre of a trunk, branch or root.

Heartwood The central section of a branch or trunk usually darker coloured than *sapwood*, comprised of *lignin* blocked *secondary xylem* sapwood cells that are no longer conductive but provide a structural and protective function from substances that form or are deposited as part of the aging process as each cell dies (Shigo 1986, p. 54).

Heat island effect Increased ambient air temperatures around buildings and hard surfaces in urban environments. This is primarily due to increased reflected or reradiated sunlight, reduced evapotranspiration and shade from trees and other vegetation, air pollution, less open bodies of water reducing evaporation and humidity, with the effect worsened when wind speed is slow.

Heave In relation to a shrinkable clay soil, expansion due to re-wetting, sometimes after felling or root severance of a tree which was previously extracting moisture from the deeper layers; also, in relation to root growth, the lifting of pavements and other structures by radial expansion; also, in relation to tree *stability*, the lifting of one side of a wind-rocked *root plate* (Lonsdale 1999, pp. 315–316).

Height Distance measured vertically between a horizontal plane at the lowest point at the base of a tree, immediately above ground, and a horizontal plane immediately above its uppermost point.

Height/diameter ratio An estimate of the ratio of trunk diameter to the *height* of the tree and applied as an alternative to *Live crown ratio*. Example, in a 20 m high tree with a trunk diameter of 500 mm, the height/diameter ratio would be 40:1 (20 ÷ 0.5 m) where a value of 60 or less is considered good. This is an indicator for the tolerance of some species to site modification especially for *stability*.

Helical Shaped like a spiral or a helix. Especially of wood fibres where the growth habit of a tree twists the fibres to resist stress loading from dominant wind flow by aligning fibres with the wind direction. See also *Thigmomorphogenesis*.

Helical crack Narrow splitting usually spiralling around a stem, internal in origin, and may continue for some distance (Mattheck & Breloer 1994, pp. 104–105).

Helical grain A wood structure in which the cells are aligned in a broad spiral pattern around the stem instead of running parallel to the stem *axis* (Lonsdale 1999, p. 316).

Heliotropism See *Phototropism*.

Hemicellulose A structural substance in cell walls comprised of chains of molecules of *sugars* other than glucose and is able to be metabolised after absorption into the cell wall, often found in association with cellulose.

Hemiparasite See *Semi-parasite*.

Heritage A country's or an area's history and historical buildings and sites that are considered to be of interest and value to present and future generations.

Heterotroph Organism that, in order to survive consumes as food the cellular materials made by another organism, e.g. a fungus does not photosynthesise but breaks down and consumes those materials produced by photosynthesis or by other organisms. See also *Autotroph*.

High vigour *Accelerated growth* of a tree due to incidental or deliberate artificial changes to its growing *environment* that are seemingly beneficial, but may result in *premature aging* or failure if the favourable conditions cease, or promote *prolonged senescence* if the favourable conditions remain, e.g. water from a leaking pipe; water and nutrients from a leaking or disrupted sewer pipe; nutrients from animal waste, a tree growing next to a chicken coop, or a stock feedlot, or a regularly used stockyard; a tree subject to a stringent watering and fertilising program; or some trees may achieve an extended lifespan from continuous *pollarding* practices over the life of the tree.

High volume deadwood Where >10 dead branches occur that may require *removal*. See also *Deadwood*, *Low volume deadwood* and *Medium volume deadwood*.

High volume dieback Where >50% of the *crown cover* has died. See also *Dieback*, *Low volume dieback* and *Medium volume dieback*.

High volume epicormic shoots Where >50% of the *crown cover* is comprised of live *epicormic shoots*. See also *Low volume epicormic shoots* and *Medium volume epicormic shoots*.

Holding wood See *Buttress wood*.

Hollow A large void initiated by a *wound* forming a *cavity* in the trunk, branches or roots and usually increased over time by *decay* or other contributing factors, e.g. fire, or fauna such as birds or insects, e.g. ants or termites. A hollow can be categorised as an *Ascending hollow* or a *Descending hollow*.

Horizontal wound Usually superficial horizontal wounding from insects burrowing between bark layers and revealed by decorticating bark. Often evident on smooth bark Eucalypts.

Hose pipe kinking *Structural failure* or *collapse* of a *hollow* stem immediately below a solid section by cross-sectional flattening caused by longitudinal splits of *neutral fibres* where more than 70% of the stem radius is *hollow* (Mattheck & Breloer 1994, pp. 36–38).

Hot spot An area on the trunk between approximately 1 m above ground and the lowest branch where *damping* is diminished or prevented due to the lack of lower *orders of branches* as a result of a high proportion of branch failures or removals by pruning occurring in the past leaving the area clear of branches and the crown section above *end loaded* (Albers & Hayes 1993; Lonsdale 1999, p. 39).

Hour glass Pronounced thinning of different regions of a palm stem indicative of *episode*s of severe environmental stress, e.g. drought or lack of nutrients, *palm over-pruning*, and may limit ongoing growth and may be potential points of weakness (Harris *et al.* 2004, p. 39).

Humectant A substance that aids in the retention or absorption of moisture.

Humic soil A soil formed mostly from *humus*.

Humus The smallest organic soil particles (Handreck & Black 2002, p. 25). It is formed from the accumulation of decayed organic matter, dark in colour and usually derived mostly from plant material after decomposition, within soil or normally arising as a surface layer to soil, e.g. animal carcasses, faeces, leaf litter, fruit and decayed wood.

Hung up branch See *Hanging branch*.

Hydraulic lift A hypothesis where roots at depth in moist soil absorb water releasing it at night to soil at a more shallow depth storing it to be reabsorbed the following day (Caldwell & Richards 1989, pp. 1–5; Craul 1992, p. 128; O'Callaghan & Lawson 1995, p. 104).

Hydraulic pressure See *Hydrostatic pressure*.

Hydraulic services plan See *Hydrology plan*.

Hydrology plan A plan showing proposed or existing drainage works.

Hydrophyte Plant adapted to survive in regularly or permanently wet soil conditions, e.g. mangroves.

Hydroponic A plant grown without soil and sustained by its roots immersed in a water and mineral nutrient solution. The roots may be supported in an inert *growing media*, e.g. perlite, gravel or mineral wool.

Hydrostatic pressure The pressure exerted upon a structure by the build-up of water at rest.

Hydrotropism A plant growth response to high moisture stimulus.

Hypha An individual strand of the branching and filamentous thread-like structures of a fungus that average 1–5 μm (microns) in diameter, masses of which form the vegetative body of the fungus called *mycelium* (Manion 1991, p. 107).

Hyphae Plural of *hypha*.

Hypocotyl Embryonic *trunk* of most trees located between the root and cotyledons (Boland *et al.* 2006, p. 690).

Hypsometer Any of several tools or instruments used to measure the *height* of a tree, e.g. a *clinometer*.

Illegal removal Tree *removal* as an action contrary to its protection by law.

IML Impulse Hammer® See *Sonic detectors.*

Immediate An *episode* or occurrence, likely to happen within a twenty-four (24) hour period, e.g. tree failure or collapse in full or part posing an imminent danger. See also *Short term*, *Medium term* and *Long term.*

Impact wound *Mechanical wound* caused by an impact *episode,* e.g. collision by a motor vehicle.

Implant A slow release fertiliser or insecticide inside a plastic capsule inserted in the *vascular cambium* of the *lower trunk* of an ailing tree. Medicap® is an example of such a proprietary system.

Impulse hammer See *Sonic detectors.*

Inappropriate tree management The planting or retention of a tree where it is known that the tree will outgrow the space available for its growth above or below ground before or at maturity, and is likely to cause disruption or damage to built structures, or retention of a tree when it is known to present a potential *hazard* to people or property. See also *Appropriate tree management*, *Tree preservation* and *Tree management.*

Incipient failure Initial *structural deterioration* of a branch, trunk or root causing the tree part or section to be deformed or cracked as a result but not collapsing or becoming detached.

Incision Wound caused by cutting or engraving. See also *Laceration*.

Included See *Included bark*.

Included bark 1. The bark on the inner side of the *branch union*, or is within a concave *crotch* that is unable to be lost from the tree and accumulates or is trapped by *acutely divergent* branches forming a *compression fork*. 2. Growth of bark at the interface of two or more branches on the inner side of a branch union or in the crotch where each branch forms a branch collar and the collars roll past one another without forming a graft where no one collar is able to subsume the other. Risk of failure is worsened in some taxa where branching is *acutely divergent* or *acutely convergent* and ascending or erect.

Included throughout *Included bark* evident on most or all *first order branches* and *lower order branches*.

Inclusion See *Included bark*.

Increment borer See *Pressler increment borer*.

Increment core The wood sample, usually 3 mm diameter, extracted by a *Pressler increment borer*.

Indigenous A *native* plant usually with a broad distribution in a particular country, geographic region or area. See also *Endemic, Locally indigenous* and *Non-locally indigenous*.

Infection Parasitic micro-organisms become established in the tissues of a host, e.g. a tree or another organism (Lonsdale 1999, p. 316).

Inferior Of *first order structural branches* (FOSB) usually growing in association with a *superior first order structural branch*, where the branch or branches have a smaller diameter and generally support a lesser percentage of crown cover than the *superior* first order structural branch often due to being suppressed because of competition for space and light. This may vary on a tree that has had its crown structure modified. See also *Apical meristem* and *Lateral*.

Inferior epicormic An epicormic shoot in a *sprout mass* that remains small due to competition and may eventually *decline*. See also *Elite*.

Infiltration rate The speed (or distance per hour) at which a column of water (height in mm) soaks into soil (Handreck & Black 2002, p. 64).

Infrastructure The basic facilities, services, and installations needed for the functioning of the community, such as roads, transportation and communications systems, water and power lines, and public institutions including schools, post offices, and prisons.

Ingrown bark See *Included bark*.

In-ground container Plant container constructed of fabric that is buried in the ground. The fabric is designed to allow small roots to escape but strangling them as they increase in diameter (Clark 2003, p. 10).

In-ground tree A tree grown in natural soil (Clark 2003, p. 11). See also *Ex-ground trees*.

Inhibiting Root barriers using chemical control or toxins to reduce root growth by applying chemicals directly into soil or painted treatments to walls or the slow release of root inhibiting chemicals impregnated into *deflecting* root barriers (Coder 1998, p. 63; Roberts *et al.* 2006, p. 355).

Inhibitor See *Root barrier* and *Inhibiting*.

Initial wound margin The site of initial wounding often evident as a faint line of discoloured bark or bark of a different texture to adjacent undamaged trunk. This may assist in determining the age of a wound.

Injection See *Tree injection*.

Injure See *Injury*.

Injury Any *immediate* or ongoing process causing wounding of a tree.

Inner bark See *Phloem*.

Inner crown The inner half of the volume of a *crown*. See also *Outer crown* and *Outer extremity of crown* and Figure 8.

Inorganic Material derived other than from living or non-living organisms, e.g. water, air and rock.

Inorganic mulch A *mulch* comprising non-cellular natural or artificial substances, e.g. stones, recycled crushed terracotta, decomposed granite. See also *Organic mulch*.

Insect wound Wounding to any part of a tree caused by insect activity, e.g. borers and termites.

In situ Occurring in its original place, e.g. soil level, *remnant vegetation*, the place from where a tree was transplanted, or where a tree is growing.

Integrated pest management (IPM) A management system that makes use of a range of different techniques in combination to control pests, concentrating on the processes least harmful to the environment and most specific to the pest, e.g. pest-resistant plant varieties, regular monitoring for pests, pesticides or promoting natural predators of the pest (Hubbard, Latt & Long 2006).

Interbuttress zone Area of trunk located between *buttress roots*.

Interception The sum of canopy/crown surface as water storage on leaves, branches, and trunk bark; and evaporation during rainfall *episode*s (Geiger 2004, p. 5) and the amount of sunlight the canopy/crown surface prevents from reaching the ground or buildings, reducing surface temperatures.

Internode The space between adjacent nodes on branches or stems separated by stem elongation.

Introduced A plant brought into a country, geographic region or area where it was not previously growing indigenously. Such plants may be associated with agriculture or horticulture as garden escapes.

Invasive root mapping Any *root mapping* process that disturbs or displaces soil or *growing media* to locate but not damage roots, e.g. *hand excavation*, or using an *air knife* or *water knife*. See also *Non-invasive root mapping*.

Invert level (IL) Level taken at the bottom of a pipe on its inside.

Iodine test The applications of iodine to a wood sample to determine the radial depth to which living cells extend. Where starch is present cells darken, or where absent cells yellow. The formula is I2-KI, known as 'Lugols' Aqueous Iodine Solution. This can be made by a chemist and consists of 10 g Iodine Crystal (Potassium iodide B.P.) in 200 mL of distilled water (Shigo 1991, p. 365).

Ion This is a positively or negatively charged atom. The removal of one or more electrons causes an atom to be positively charged and the addition of one or more electrons causes an atom to be negatively charged (Handreck & Black 2002, p. 16).

Irreversible decline The decline of a tree where it has progressively deteriorated to a point where no remedial works will be sufficient to prevent its demise, usually of *poor form* and *low vigour*. See also *Spiral of decline*.

Isolated tree 1. A tree growing as a solitary specimen in an exposed location away from other trees as a result of natural or artificial causes and may be naturally occurring. 2. A tree planted as a solitary specimen in an exposed location away from other trees. Trees that become isolated as a result of changes in their growing environment may adjust over time to survive or may decline or succumb to the problems of exposure, e.g. *windthrow*. A planted isolated tree will usually be *open grown*, and hence strong, with generally a lower height and broad spreading crown as a result of a lack of competition. See also *Height/Diameter ratio*.

Judge An individual empowered to critically examine facts and *evidence* and law for cases based on law and make judgments on matters other than law.

Juvenile See *Young*.

Kerb The usually concrete or stone, upright edge, at the outer edge of a formed roadway to retain the soil in the road reserve, guide drainage of the roadway and storm water pipes from adjoining properties.

Kill zone Regular pruning of roots or chemical treatment to soil at a given distance from a tree, to restrict root growth near other plants, usually applied to field crops (Coder 1998, p. 61).

Kino The extractive polyphenols (tannins) formed in veins in the *cambial zone* as a defence in response to wounding in eucalypts. Often visible as an *exudate* when the kino veins rupture or are injured (Boland *et al.* 2006, p. 691).

Knees See *Pneumatophore*.

Krummholtz See *Krummholtz form*.

Krummholtz form Trees with a prostrate codominant form due to wind and snow loading in the sub arctic and near the tree line on mountains (Ennos 2001, p. 44). See also *Flagging*.

Laceration Wound caused by tearing. See also *Incision*.

Laminated joint Successive layers of branch wood increments intertwining with branch/trunk wood forming an optimal union. See also *Delaminated joint*.

Land and environment court Courts specifically dealing with environmental and planning issues, e.g. in New South Wales, Australia, the Land and Environment Court.

Landscape plan A plan to depict landscaping works.

Landscape planting New trees, shrubs and ground covers planted into the landscape.

Large deadwood A dead branch >10 mm diameter and usually >2 m long, generally considered of high *risk* potential. See also *Deadwood* and *Small deadwood*.

Large tree A tree with a height >20 m or *crown spread* >20 m at maturity, *in situ*. See also *Size of tree*, *Small tree* and *Medium tree*.

Last season's growth The *growth increment* formed in the growing season prior to the current growing period. See also *Current season's growth* and *Growth rings*.

Latent bud Concealed bud more than one year old that has grown enough each year so that its growth is at or near the surface of the bark where it may develop normally under suitable environmental conditions.

Lateral Usually an *inferior* branch growing from the side of a dominant branch.

Lateral bud A bud forming to the side of an *apical bud* that gives rise to branches or stems that are secondary growing away from the *apical bud*.

Lateral leeward A lateral structural first order root occurring on the downwind side of a tree, subject to *compression* as the trunk and crown are affected by subsequent loading. See also *Lateral windward*.

Lateral pruning Pruning to restrict or reduce *crown spread* horizontally.

Lateral roots The branching of a root horizontal to its orientation through the differentiation of *parenchyma* cells in the endodermis. Such roots form from the *radicle* and may grow to replace the *radicle* or may persist to become woody structural roots. Lateral roots may form near the *root tip* behind the *root hairs*.

Lateral windward A lateral structural first order root occurring on the upwind side of a tree, subject to *tension* as the trunk and crown are affected by subsequent loading. See also *Lateral leeward*.

Late season's wood In some trees, small cells with few vessels but with thick walls produced as a consequence of slow growth as day length and temperature decreases during autumn and winter. See also *Early season's wood*.

Late wood See *Late season's wood*.

Latex The extractive substance contained within laticiferous vessels and tubes. Latex is usually white in colour and is present in many trees such as *Ficus* spp.

Layback The section of the kerb angled back away from the roadway to allow for vehicular access across the kerb and gutter and used to construct the *vehicular crossover* to the adjoining property for a driveway.

Leader A structural branch asserting apical dominance.

Leaf area index (LAI) The percent of *crown projection* covered by foliage (Geiger 2004, p. 2).

Leaf area indices The plural of *Leaf area index*.

Leaf density See *Crown density*.

Leaf litter The accumulation of fallen leaves, e.g. on the ground beneath plants, in branch crotches or gutters.

Leaf scar Small wound-like protrusion remaining after the abscission of a leaf.

Lean See *Leaning*.

Leaning A tree where the *trunk* grows or moves away from upright. A lean may occur anywhere along the *trunk* influenced by a number of contributing factors, e.g. genetically predetermined characteristics, competition for space or light, prevailing winds, aspect, slope, or other factors. A *leaning* tree may maintain a *static lean* or display an increasingly *progressive lean* over time and may be hazardous and prone to *failure* and *collapse*. The degrees of leaning can be categorised as *Slightly leaning, Moderately leaning, Severely leaning* and *Critically leaning* (Figure 19).

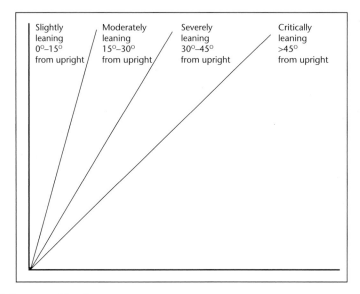

Figure 19 Degrees of leaning.

Leaning tree See *Leaning*.

Lee See *Leeward*.

Leeward The side of an object opposite to that exposed to wind direction or the underside of a lean. See also *Windward*.

Legislation Laws made by a Parliament or Congress.

Lenticel Pore cell within outer layers of a woody stem providing for the exchange of gases between the plant and the atmosphere.

Lesion Any dead spot on living tissue (Shigo 1991, p. 140). See also *Canker*.

Levels of government Tiers of government to assess national, regional or local issues, e.g. Federal, State and Local Governments respectively.

Lichen The symbiotic association between an alga and a fungus, often evident in a forest environment where it grows on trees without harming them. Lichen growth is in the forms known as foliose, fruticose and crustose (Manion 1991, p. 104).

Life cycle Progressive sequences a tree undergoes from fertilisation to death including the production of *progeny* through the stages of *Young*, *Mature* and *Over-mature*.

Life expectancy The estimated life span, or remaining life span of a tree.

Lightning strike wound A wound from a lightning strike. Such a wound may kill a tree outright or cause it to catch fire, or may destroy the tree in full or part, or no injury may be evident and a tree gradually declines through resulting *stress*. Bark may be exploded from the tree by pressure radiating from the core of the lightning path resulting in further compounded damage through water heating and steam explosions in the tissues and the electrical disruption of living cells (Coder 2004, pp. 35–44).

Lignification The gradual deposition of *lignin* within cell walls, providing strength and support to wood. See also *Delignification*.

Lignin A durable chemical component deposited in aging cell walls comprising between 18–35% of wood (Thomas 2000, p. 41), providing strengthening against *compression* and *tension* forces.

Lignotuber A woody tuber developed in the axils of the cotyledons and the first few leaf pairs becoming massive in some mature trees of mallee Eucalypt, evident as *gall* like swellings at the base of some trees (Brooker & Kleinig 1999, p. 341).

Likely habitat tree See *Potential habitat tree.*

Limb See *Branch.*

Line clearance Pruning to maintain clearances around aerial utility cables.

Lintel A usually short elongated prefabricated load-bearing structure located horizontally to span short distances between piers or footings.

Lion's tailing The result of incorrect pruning where foliage is concentrated at the distal ends of branches. See also *End loading.*

Lipping (Lonsdale, 1999, p. 317), see *Crotch seam.*

Live crown ratio (LCR) An estimate of the ratio of the length of live crown to the *height* of the tree and usually applied to conifers. This is usually expressed as a percentage, e.g. 20 m high tree with a live crown of 10 m long, the LCR would be 50% ($10 \div 20$). LCR is an indicator for the tolerance of some species to site modification.

Load See *Loading.*

Loading Weight that is carried, e.g. as bending *stress* on a *branch*. See also *Static load* and *Dynamic load.*

Local environmental plan (LEP) A *statutory plan* usually of broader significance and control prepared under the requirements of *Legislation* that controls development at the local government level.

Locally indigenous A *native* plant as *remnant vegetation, self-sown* or planted in an area or region where it occurred originally. See also *Endemic, Indigenous* and *Non-locally indigenous.*

Longevity Long lived, referring to a plant living for a long period of time.

Longitudinal rib A rib formed usually over a radial crack in a stem, e.g. trunk (Mattheck and Breloer 1994, p. 104). See also *Rib, Round-edged rib* and *Sharp-edged rib.*

Long term A period of time greater than >40 years. See also *Periods of time, Immediate*, *Medium term* and *Short term*.

Loosened bark Localised area of a stem subject to *tension* causing bark to become spread out, segmented or detached (Mattheck & Breloer 1994, p. 174). See also *Congested bark*.

Loose tipping On clay capped landfill sites the concept of working from the capped surface with plant equipment to level introduced soil or *growing media* for new plantings to reduce the incidence of compaction (Roberts *et al.* 2006, p. 106; Dobson & Moffat 1993).

Lop See *Lopping.*

Lopping Cutting between branch unions (not to *branch collars*), or at inter-nodes on a *young* tree, with the *final cut* leaving a *stub or palm over-pruning*.

Low canopy *Crowns* of more than one tree or a *stand* of trees where the branches and foliage extend close to the ground.

Low crown The *crown* of a tree where the branches and foliage extend close to the ground.

Low vigour Reduced ability of a tree to sustain its life processes. This may be evident by the *atypical* growth of leaves, reduced crown cover and reduced crown density, branches, roots and trunk, and a deterioration of their functions with reduced resistance to predation. This is independent of the condition of a tree but may impact upon it, and especially the ability of a tree to sustain itself against predation. See also *Vigour, Normal vigour* and *High vigour*.

Low volume deadwood Where <5 dead branches occur that may require *removal*. See also *Deadwood, Medium volume deadwood* and *High volume deadwood*.

Low volume dieback Where <10% of the *crown cover* has died. See also *Dieback, High volume dieback* and *Medium volume dieback*.

Low volume epicormic shoots Where <10% of the *crown cover* is comprised of live *epicormic shoots*. See also *High volume epicormic shoots* and *Medium volume epicormic shoots*.

Lower crown The *proximal* or lowest section of a crown when divided vertically into one-third (⅓) increments. See also *Crown*, *Mid crown* and *Upper crown*.

Lower order roots Orders of roots other than first order roots, e.g. second order, third order, fourth order etc.

Lower order branches Orders of branches other than first order branches, e.g. second order, third order, fourth order etc.

Lower trunk Lowest, or *proximal* section of a trunk when divided into one-third (⅓) increments along its *axis*. See also *Trunk*, *Mid trunk* and *Upper trunk*.

Macroclimate The conditions of climate extending over a large geographical area. See also *Climate*, *Mesoclimate* and *Microclimate*.

Macronutrients Essential soil elements in solution derived from mineral solids required in large amounts for the growth of plants, Nitrogen, Phosphorus, Potassium, Calcium, Magnesium, Sulfur (Sulphur). See also *Essential elements* and *Micronutrients*.

Macropores Large *pore* spaces between soil aggregates or between *ped*s. See also *Pore*, *Mesopores* and *Micropores*.

Maiden tree A tree that has never been altered by pruning.

Maintenance Any works including pruning, weeding, mulching, fertilising and watering undertaken to prolong the *vigour* and life expectancy of a tree. See also *Crown maintenance*, *Planned maintenance* and *Unplanned maintenance*.

Mallee A shrub to small tree of eucalypts with a *crown* formed from multiple stems, often subject to fire where the crown is destroyed and regenerates as a *coppice* from a *lignotuber* (Beard 1990, p. 128).

Managed decline The concept of undertaking an ongoing program of maintenance works on a declining tree to prolong retention over its remaining life, often associated with a *significant tree*. See also *Cost analysis*.

Marlock Single stemmed mallee species of eucalypt from Western Australia that after fire regenerates only from seed due to the absence of a *lignotuber* (Beard 1990, p. 128).

Mast year A year when a tree produces an abundance of fruit uncharacteristic-ally greater than in previous years potentially linked to broader environmen-tal cycles (Suzuki & Grady 2004, p. 110).

Mature Tree aged 20–80% of life expectancy, *in situ*. See also *Age, Young* and *Over-mature*.

Mechanical wound Wounding inflicted by abrasion, e.g. by motor vehicles, grass mowing equipment, grazing by horses, cows or birds (parrots); impact, e.g. by motor vehicle collisions; drilling, e.g. with increment cores, resisto-graphs, cable bracing, hanging pots, hammocks etc.; branch tearing, e.g. from wind damage, collision from falling branches, vandalism; and root sev-erance, e.g. root pruning for excavation for building or utility services or for agricultural cultivation.

Media See *Growing media*.

Medium See *Growing media*.

Medium frass *Frass*, grain-like to the touch with particles ranging from 0.1–1 mm in diameter. See also *Frass, Coarse frass* and *Fine frass*.

Medium term A period of time 15–40 years. See also *Periods of time, Immediate, Short term* and *Long term*.

Medium tree A tree with a height of 10–20 m or *crown spread* of 10–20 m at maturity, *in situ*. See also *Size of tree, Small tree* and *Large tree*.

Medium volume deadwood Where 5–10 dead branches occur that may require *removal*. See also *Deadwood, High volume deadwood* and *Low volume deadwood*.

Medium volume dieback Where 10–50% of the *crown cover* has died. See also *Dieback, High volume dieback* and *Low volume dieback*.

Medium volume epicormic shoots Where 10–50% of the *crown cover* is com-prised of live *epicormic shoots*. See also *High volume epicormic shoots* and *Low volume epicormic shoots*.

Medullary rays See *Vascular ray.*

Meristem See *Apical meristem.*

Meristematic A cellular area containing actively dividing or potentially actively dividing cells, e.g. apical buds, axillary buds and roots (Bailey 1999, p. 293).

Mesoclimate The conditions of climate extending over a small area or district, e.g. a city, valley or low lying area. See also *Climate*, *Microclimate* and *Mesoclimate.*

Mesophyte A plant adapted to survive in moderately moist soil conditions.

Mesopores Medium *pore* spaces in soil containing water available to plants. See also *Pore*, *Micropores* and *Macropores.*

Metriguard Stress-wave Timer® See *Sonic detectors.*

Microaerophilic Relating to biochemical processes or organisms adapted to an environment low in oxygen (Lonsdale 1999, p. 317).

Microclimate The conditions of climate extending over a very small area, e.g. beneath the crown of a tree or physical *environment* immediately influencing a tree. See also *Climate*, *Macroclimate* and *Mesoclimate.*

Microfibrils Smallest component units of a cellulose filament in a plant cell wall (Lonsdale 1999, p. 317).

Micronutrients Essential soil elements in solution derived from mineral solids required in small amounts for the growth of plants, Iron, Manganese, Boron, Molybdenum, Copper, Zinc, Chlorine and Cobalt. See also *Essential elements* and *Macronutrients.*

Micropores Small *pore* spaces between soil aggregates of a *ped* containing water unavailable to plants. See also *Pore*, *Mesopores* and *Macropores.*

Mid crown The middle section of a crown when divided vertically into one-third (⅓) increments. See also *Crown*, *Lower crown* and *Upper crown.*

Mid trunk A middle section of a trunk when divided into one-third (⅓) increments along its *axis*. See also *Trunk*, *Lower trunk* and *Upper trunk.*

Misshapen *Atypical habit* or disfigured shape of a tree or tree part caused by wounding, decay, pruning, injury, wind, insect damage, loading from snow or from restrictions to its requirements for space or light or topography such as sloping ground.

Mistletoe *Parasitic* and *epiphytic evergreen angiosperms* that grow on the stems of trees by the use of cell structures called *haustoria*, consuming nutrients and water produced by the host but most produce their own *sugars* by photosynthesis. The fruit are spread in the faeces of fauna that deposit the seeds on stems or in the crotches of small branches or the mucilaginous fruit may be spread by sticking to fauna. The effects of an infestation may cause reduced *vigour*, modify structure and contribute to the *decline* of a tree (Coder 2004, pp. 37–44). Of the 86 known species of mistletoe occurring in Australia, most are host specific (Reid 1996, p. 2).

Mitigation 1. See *Remedial pruning.* 2. Any action taken to reduce *risk,* e.g. *pruning* or *removal.*

Mixed age population A population of trees that contains representatives of each age class, i.e. *young, mature* and *over-mature,* so that trees in the population will not all mature or become *senescent* at the same time. This is usually achieved by staggered planting and/or removal of trees in a systematic manner. This also spreads the cost of planting, maintenance and removal over time.

Moderately crooked Stem section growing at a lean 30–60° from upright, *atypical* from its natural habit or *crown form.* See also *Crooked, Slightly crooked* and *Severely crooked.*

Moderately leaning A leaning tree where the trunk is growing at an angle within 15–30° from upright. See also *Leaning, Slightly leaning, Severely leaning* and *Critically leaning.*

Monocot See *Monocotyledon.*

Monocotyledon The single embryonic seed leaf of some *angiosperms* that gives rise to palm trees and many other plants, e.g. grasses. See also *Cotyledon* and *Dicotyledon.*

Monoculture The cultivation of one type of plant only, usually in reference to crops.

Moribund Advanced state of decline, dying or nearly dead.

Mortality spiral See *Spiral of decline*.

Mud/mudguts Of termites, soil and excrement from digested wood.

Mulch Any *organic* or *inorganic* material placed over the soil or growing media near cultivated plants primarily to suppress weeds, modify soil temperature and retain soil moisture levels. This artificial process generally imitates the function of *leaf litter* in a forest environment. See also *Inorganic mulch*, *Organic mulch* and *Species specific mulch*.

Multiple first order branches Two or more first order branches arising from a trunk. Such may be *dual leader branches* or *multiple leader branches*, arising initially from a division at the termination of the trunk, generally *divergent* and *ascending* from vertical tending to upright at a sufficient distance from each other and can be categorised as *temporary branches*, *permanent branches*, *superior* or *inferior*.

Multiple leader branches A crown comprised of three or more *codominant first order structural branches (FOSB)*, where two or more of the branches have a similar diameter or support an approximately even percentage and volume of *crown cover*.

Multi-trunked See *Acaulescent* and *Trunk*.

Mycelia Plural of *mycelium*.

Mycelium The vegetative part of a fungus comprised of a mass of *hyphae*.

Mycology The scientific study of *fungi*.

Mycorrhiza A combination of fungus and plant roots forming a symbiotic or weakly pathogenic association. There are a number of types: ectomycorrhiza, where the fungus is on the surface of the roots and forms a Hartig net; endomycorrhiza, where the fungus invades the roots where it is frequently confined to well defined layers; ectendomycorrhiza, where there is a Hartig net but the fungus also penetrates the roots; and vesicular arbuscular mycorrhiza (VAM), where the hyphae in the cortical cells of the root may be coiled or branched (Grgurinovic 1997, p. 701).

Mycorrhizae Plural of *mycorrhiza*.

Mycorrhizal association Fungus and plant roots forming a symbiotic or weakly
pathogenic association where intercellular penetration of the *epidermis* by
hyphae (Craul 1992, p. 52) increases significantly the number of short roots
and surface area for the absorption of moisture and more efficient absorbing
of mineral elements particularly nitrogen and phosphorous. Increased efficiency is provided for the root by the capacity of the fungus to make available extracted elements from organic matter in the soil. As the fungus
parasitises the root it receives nutrients and carbohydrates, while forming a
barrier in the root against other pathogens (Manion 1991, p. 129).

Native A plant found to occur as an *endemic* or *indigenous* species where it is growing or a plant known to have originated as an *endemic* or *indigenous* species from a particular place, e.g. continent, country, region, mountain or island.

Nature strip Sometimes a name given to the entire *road reserve* but more broadly used to refer to the usually narrow *planting strip* section of the road reserve.

Naturalised A plant introduced from another country or region to a place where it was not previously *indigenous* where it has escaped from agriculture or horticulture or as a garden *escape* and has sustained itself unassisted and given rise to successive generations of viable *progeny*.

Natural pruning Shedding of branches, usually through their *compartmental-isation* after injury or from overshadowing. This can occur when a branch becomes inefficient as a source of photosynthates, progressively becoming thinned of foliage, eventually dying off being ultimately shed from the tree often after it has decayed, and breaking near the branch collar.

Natural target cut See *Final cut*.

Natural target pruning See *Final cut.*

Necrotic Dead area of tissue that may be localised, e.g. on leaves, branches, bark or roots.

Negligence With regard to trees, failure to take reasonable care to prevent hazardous situations from occurring which may result in injury to people or damage to property (Lonsdale 1999, p. 317). See also *Nonfeasance.*

Nest Any structure built or naturally formed on, from or within a tree to support as habitat or shelter any part of the life cycle of fauna.

Nesting box Artificial structure, usually a box and of varying dimensions to suit individual requirements of different species of birds and mammals, located in trees or other structures to mimic nesting requirements and often used where a tree with a *nesting hollow* is to be removed or where the *hollow* has been lost, e.g. by fire and nesting by honey bees or feral birds (Grant 1997, pp. 4–7).

Nesting hollow A *hollow or cavity* within any part of a tree utilised as habitat or shelter for any part of the life cycle of fauna, e.g. birds, reptiles or mammals.

Neutral fibres Where fibres subject to bending exhibit neither *tension* or *compression* nor shortening or elongation (Mattheck 2004, p. 17). Such fibres are often located towards the sides of a bend in a trunk or branch.

New season's growth See *Current season's growth.*

Nodal Of, or being like a node, or positioned at or near a node.

Node Raised point on a stem from where one or more leaves or buds arise, or have arisen.

Nonfeasance A failure to carry out an official duty or legal requirement. See also *Feasance.*

Non-grafted union Roots or branches growing together without forming a graft from the same tree where they were convergent, or crossed, or entwined. This will be common on trees of different taxa where grafting is unlikely to occur. Such a graft on the same tree or of same or similar taxa may be prevented by *persistent bark* or where no abrasion of persistent bark was evident and where movement between the two structures persists. The roots of a strangler fig form a non-grafted union with their host while it is alive.

Non-invasive root mapping Any *root mapping* process that does not disturb or displace soil or *growing media* to locate roots, e.g. *ground penetrating radar.* See also *Invasive root mapping.*

Non-locally indigenous A *native* plant, *self-sown* or planted in an area or region where it originally did not occur. See also *Endemic, Indigenous* and *Locally indigenous.*

Non-structural branches Usually *first order branches* arising from the trunk and sometimes *lower order branches* that are ancillary and do not form a structural framework of branches supporting the *crown*. Such branches are often temporary but may persist beyond the tree's maturity or be shed by *natural pruning*. On trees of *forest form* or where dense shade is cast to the underside of the *foliage crown*, such first order branches may be encountered arising radially with each *inferior* and usually as *temporary branches*, *divergent* and ranging from horizontal to ascending, often with *internodes* exaggerated due to competition for light and space. These branches may occur on a tree in a forest environment as lower branching where trunk elongation is evident to elevate the *crown* up through the *canopy*. These may also be *epicormic shoots* as *watershoots, suckers* or *adventitious shoots* that have arisen after an *episode* of stress or as a seasonal characteristic of a species, e.g. *Jacaranda mimosifolia*, and may form *temporary branches* or persist as *permanent branches.*

Non-woody roots See *Fine roots.*

Normal vigour Ability of a tree to maintain and sustain its life processes. This may be evident by the *typical* growth of leaves, crown cover and crown density, branches, roots and trunk and resistance to predation. This is independent of the condition of a tree but may impact upon it, and especially the ability of a tree to sustain itself against predation. See also *Vigour, Low vigour* and *High vigour.*

Notch A *cavity* or *hollow,* e.g. from a wound or where the remains of a branch has been isolated within an *occluded* section of stem, or decayed leaving a void. See also *Notch stress.*

Notch stress Any wound, *cavity* or *hollow* on a tree that disrupts the uniform loading of stress. See also *Axiom of uniform stress* and *Notch.*

Notional See *Notional defect.*

Notional defect A *structural defect* of *decay* or *cavity* that has not been confirmed as either symmetrical or asymmetrical within a stem or root (Mattheck & Breloer 1994, p. 185).

Noxious weed A plant species of any taxa declared a weed by legislation. Treatment for the control or eradication of such weeds is usually prescribed by the legislation.

Nuisance The real or perceived loss of use and enjoyment of land through the growth of a tree or its branches, roots, leaf or fruit fall onto another property.

Nurse crop Plantings of trees, shrubs or grasses to provide shelter from wind, suppress weeds, and shade the soil surface to assist in the establishment of a desired crop or plantings. Such plantings may be with taxa of low competitiveness or sparse plantings with competitive taxa at times that minimise competition with the main crop or plantings (Hitchmough 1994, p. 142; Robinson 2004, p. 209).

Nurse planting See *Nurse crop.*

Nurse tree Plantings of fast growing trees as a *nurse crop*, often of *pioneer tree* species, to protect and shelter slow growing and sometimes smaller trees during their establishment. Such trees are often short lived and likely to be removed or have numbers reduced after the desired usually long-lived trees are established, or decline naturally when the slower trees become established and outcompete the nurse trees (e.g. they become *overtopped*) (Loudon 1853, p. 440).

Nutrients See *Essential elements, Macronutrients and Micronutrients.*

Objectors Individuals or parties opposing the proposed use or development of a site.

Oblique roots See *Heart roots.*

Obtuse branch crotch A branch crotch where the angle on the inner side of the union is greater than >90°. See also *Acute branch crotch.*

Obtusely convergent Branch growing in a direction towards its point of attachment where the angle in the crotch is greater than >90° and less than <180°.

Obtusely divergent Branch growing in a direction away from its point of attachment where the angle in the crotch is greater than >90° and less than <180°.

Obvert level (OL) Level taken at the top of a pipe on its inside.

Occluded *Wound wood* growth that has enclosed the *wound face* by the process of *occlusion.*

Occluding tissue The woody tissue forming around the perimeter of a wound being a succession of *callus wood, wound wood* and *wood.*

Occlusion Growth processes where *wound wood* develops to enclose the *wound face* by the merging of *wound margins* concealing the *wound* and restoring the growing surface of the structure with each *growth increment* gradually realigning *fibres* in the wood longitudinally along the stem to maximise uniform stress loading. See also *Axiom of uniform stress*.

Occlusion seam A line of *included bark* where the interface of merging *wound margins* is *occluded* or forms a *partial occlusion*.

Occupancy rating The frequency of use of a likely target and possibility that people will be present when tree failure or collapse occurs.

Offspring See *Progeny*.

Old See *Over-mature*.

Old tree See *Over-mature*.

Open grown A tree with *crown form dominant*, grown singly without competition for space and light, exposed to wind, often resulting in a shorter tree with a broad spreading crown that extends towards the ground (Matheny & Clark 1998, p. 18) (Figure 20). See also *Forest grown*.

Open wound Wound with poor to non-existent perimeter or *callus wood* or *wound wood* on an older wound without well-defined apex, base or margins and often this will be associated with a recent wounding *episode* or an older

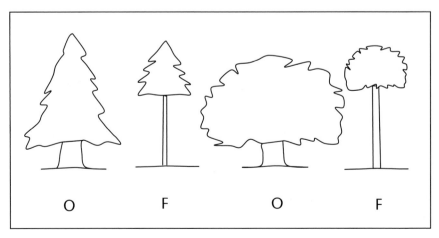

Figure 20 Examples of trees that are open grown (O) and forest grown (F). (Source: Matheny & Clark 1998)

episode on a senescent tree or a tree in *poor condition* or of *low vigour*, or where repeated wounding episodes such as inflicted by ongoing borer activity damages and continually alters wound perimeters, or repeated scalping of exposed roots by lawn mowing equipment.

Orders of branches The marked divisions between successively smaller branches (James 2003, p. 168) commencing at the initial division where the trunk terminates on a *deliquescent* tree or from *lateral* branches on an *excurrent* tree. Successive branching is generally characterised by a gradual reduction in branch diameters at each division, and each gradation from the trunk can be categorised numerically, e.g. first order, second order, third order etc. (See Figure 21.)

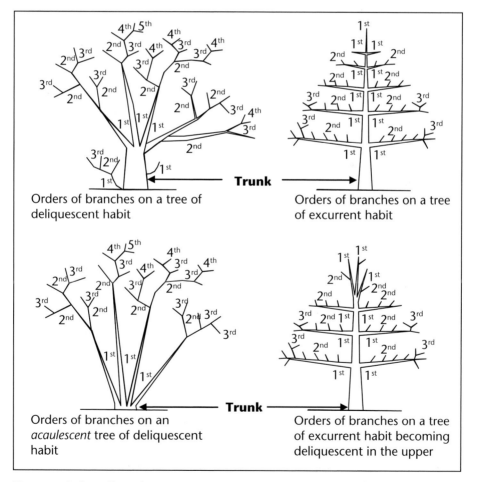

Orders of branches on a tree of deliquescent habit

Orders of branches on a tree of excurrent habit

Orders of branches on an *acaulescent* tree of deliquescent habit

Orders of branches on a tree of excurrent habit becoming deliquescent in the upper

Figure 21 Orders of branches.

Orders of roots The marked divisions between woody roots, commencing at the initial division from the base of the trunk, at the *root crown* where successive branching is generally characterised by a gradual reduction in root diameters and each gradation from the trunk and can be categorised numerically, e.g. *first order roots*, second order roots, third order roots etc. Roots may not always be evident at the *root crown* and this may be dependent on species, age class and the growing environment. Palms at maturity may form an adventitious root mass. (See Figure 22.)

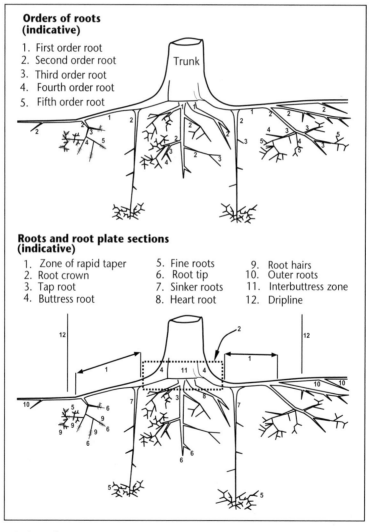

Figure 22 Orders of roots.

Ordinance The prescribed rules of a *planning instrument,* e.g. tree management order.

Organic Materials containing carbon, in combination with other elements (notably hydrogen, oxygen, nitrogen and sulfur), and that are or were parts of plants or animals (Handreck & Black 2002, p. 22).

Organic matter (OM) See *Soil organic matter.*

Organic mulch *Mulch* comprised of plant byproducts, and occasionally animal manures, e.g. chipped wood and leaves, straw or compost. See also *Inorganic Mulch.*

Ornamental tree Tree grown on private or public land for its *crown form*, colour, flowers, fruit, shade or purposes other than for the commercial production of wood, paper, oil, flowers or fruit products.

Orthotropism Growth direction taken by a plant or plant part along a vertical *axis* that is directly towards, or directly away from the source of a stimulus.

Outer bark See *Phellem.*

Outer crown The outer half of the volume of a *crown*. See also *Inner crown, Outer extremity of crown* and Figure 9.

Outer extremity of crown The outer surface area of the volume of a crown. See also *Inner crown, Outer crown* and Figure 9.

Outer roots The roots generally found beyond the *dripline* that extend with only gradual taper and are usually of small diameter 20–50 mm and opportunistically extend to seek water (Thomas 2000, p. 76).

Overgrowth See *Occlusion.*

Over-mature Tree aged greater than >80% of life expectancy, *in situ,* or *senescent* with or without reduced *vigour*, and declining gradually or rapidly but irreversibly to death. See also *Age, Young* and *Mature.*

Over-thinning The removal of a disproportionate amount of *lateral* branches and associated live crown by pruning.

Overtopped Crown of a tree restricted for light from above by the growth of another tree/s. See also *Crown form suppressed*.

Over excavation Excessive excavation beyond the required construction dimensions usually to allow safe working access or to create *benching* where the *profile* may be deep or unstable. The impact of such works must be considered where a structure is to be located near a protected tree.

Palm A *monocotyledonous angiosperm tree* that may be of single stem, or clumping, rarely branched, have palmate or pinnate leaves (fronds) and is found from tropical to subtropical regions. The trunk and roots of most palms experience only *primary thickening* and have *vascular cambium* in *diffuse bundles*.

Palm over-pruning *Lopping* to remove dead and live fronds where the removal of the live fronds damages the apical meristems as they grow in the region that tapers away from the growing point in most palms, disrupting stem thickening and exposing the tree to pathogens (Shigo 1991, p. 316).

Palm pruning Maintenance of a palm *crown* by removal of dead fronds only, protecting the growth tip and live fronds growing from the meristematic tissue that tapers away from the growing point in most palms, and the trunk that forms as those cells mature and grow larger (Shigo 1991, p. 316).

Palm topping See *Palm over-pruning.*

Parasite An organism living on or in another living organism (host) and acquiring its food from the host often to the detriment of the latter.

Parasitic See *Parasite.*

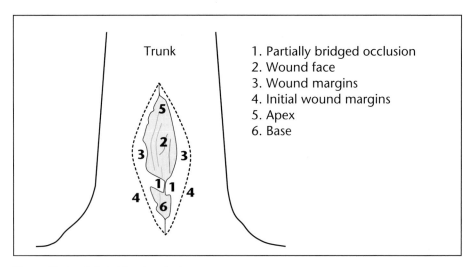

Figure 23 Partially bridged occlusion.

Parenchyma The relatively unspecialised tissue within higher plants, including the network of living cells within *sapwood* as axial parenchyma arranged vertically and ray parenchyma forming *medullary rays* (Lonsdale 1999, p. 318).

Partial occlusion *Wound wood* growth that encloses some of the *wound face* by the merging and *grafting* of some sections of the *wound margins*. Usually evident by reduced *wound face* width and indicated where an *apex* or *base* is *acute* with the vertical extent often indicated by the length of an *occlusion seam*.

Partially bridged occlusion *Wound wood* partly forming an *occlusion* by joining areas of the *wound margins* across the *wound face* at point/s other than the base or apex and may form an *occlusion seam* (Figure 23).

Partially delaminated A *mechanical wound* caused when the bark is partly stripped from a tree, usually from the trunk as a continuous sheet back to the *vascular cambium*, but remains connected with living tissue evident as *callus wood* and then *wound wood* at the resulting *wound margin*. *Scribing* should be undertaken to promote *occlusion*. See also *Delaminate*.

Pathogen Any organism causing disease, e.g. fungus, bacterium, mycoplasma, virus, viriod, nematode or protozoan.

Ped A unit of *soil structure* formed from the aggregation of soil particles into approximate geometric patterns (Craul 1992, p. 18).

Pedal *Soil structure* where aggregates within *soil horizons* are formed into crumb-like clusters or *peds*.

Peg root See *Pneumatophore*.

Pencilling Pronounced thinning of a region of a palm stem below the crown indicative of an *episode* of severe environmental stress, e.g. drought or lack of nutrients, and may limit ongoing growth and may be a potential point of weakness (Harris *et al.* 2004, p. 39).

Penetrometer A self powered non-destructive drilling device that operates a thin spade drill bit at a constant forward speed (Mattheck & Breleor 1993, p. 191), to record soundness of wood by measuring resistance to drilling that is plotted on graph paper or by a computer. Examples of such proprietary devices are the Resistograph®, Densitomat® and Sibert® (Nicolotti & Miglietta 1998, pp. 297–302).

Perched water table A localised area of saturated soil above the normal *ground water level*. This can be caused by the localised presence of relatively impervious soil layers, or soil layers disconnected by impervious rock.

Perching Tree or part thereof preferred as a place of resting or point of observation for birds.

Perching tree See *Perching*.

Perennial A plant living for more than two growing seasons.

Perennial canker See *Canker*.

Periderm In bark a protective tissue of corky suberinised cells often produced in response to wounding, or in the stem or root cortex. With secondary thickened stems it replaces the *epidermis* which splits and peels away (Rudall 1992, p. 29). It consists collectively of the *phellogen, phellem* and *phelloderm*.

Periods of time The life span of a tree in the urban environment may often be reduced by the influences of encroachment and the dynamics of the environment and can be categorised as *immediate, short term, medium term* and *long term.*

Permanent branches The main framework of *structural branches* supporting the *crown*, usually developed by the time a tree has reached its *mature* form, primarily comprised of *first order branches* and to a lesser extent some lower order branches. See also *Non-structural branches* and *Temporary branches.*

Permanent root A root that may live for a long period of time before declining or dying back, being replaced by new roots branching nearby or may persist for the life of the tree such as structural *lateral roots,* e.g. first order roots.

Permanent wilting point Permanent loss of *turgor pressure* in plant cells from which the plant will not recover. See also *Wilting, Wilting point* and *Temporary wilting point.*

Permeability See *Infiltration rate.*

Persistent bark Bark not shed annually (Boland *et al.* 2006, p. 695) and retained by a tree over many seasons' growth and shed from the outer edge of the bark very slowly, e.g. *Quercus suber* – Cork Oak or *Melaleuca quinquen-ervia* – Broad-leaved Paperbark. This may also occur on wounds concealing the extent of damage or predators such as borers. See also *Decorticate.*

pH A measure of Hydrogen *ion*s in solution and their concentration, where a value of 0 (zero) is highly acid, a value of 14 is highly alkaline and 7 is a neutral value (neither acidic nor alkaline).

Phased target pruning Specific pruning intervals where selectively, new shoots are pruned over time to restore a natural crown shape or desired *crown form* or to prevent worsening defects to reduce the risk of hazard.

Phellem In bark these cells give rise to corky tissue, are tightly packed and have deposits of *suberin* in their walls to form an impervious layer to prevent water loss and to protect against injury (Rudall 1992, p. 29). In some trees this is the initial layer of bark outside of the *epidermis.*

Phelloderm Cells formed in bark which is non-suberinised and parenchymat-ous as part of the secondary cortex (Rudall 1992, p. 29). These cells are distinguished from those of the cortex by their arrangement in radial columns reflecting their origin from the *phellogen*. As most trees age this does not keep pace with the radial growth of the stem and ruptures giving rise in some trees to rough and fissured bark.

Phellogen In bark a single layer of thin walled cells that gives rise to the *phellem* to the outside and the *phelloderm* to the inside (Rudall 1992, p. 29).

Phloem Vascular tissue with the primary role of conducting *sugars* and other nutrients and is usually external on a stem to *xylem* being formed at the outer side of the *vascular cambium* as *primary phloem*, then *secondary phloem* and the site where bark originates.

Photo montage A photographic representation of the before and after effects of *development* impact.

Photosynthesis In most green plants the process that converts light *energy* into chemical energy, with the uptake of carbon dioxide and production of water as a bi-product.

Phototropism A directional growth movement towards light (positive tropism) or away from a source of light (negative tropism, *Aphototropic*).

Phreatic level Soil and rocks beneath *underground water.*

Phreatophyte Tree with a long taproot (Russell & Cutler 2003, p. 252).

Phytopathology Scientific study of plant diseases that are caused by pathogens and environmental conditions.

Phytotelmata Pooling of water in parts of plants, e.g. in *cavities, hollows, pocket crotch* or between *anastomosing* roots (Clarke 1997, pp. 39–40).

Phytotoxic Poisonous to plants.

Phytotoxin Any substance poisonous to plants, or poisonous substance produced by a plant.

Picus® tomograph See *Ultrasonic detectors.*

Pier A vertical prefabricated support as an elevated *footing* for a built structure that is used to span over a void usually to another such structure.

Pier and beam Construction combining piers and beams to minimise soil disturbance for the footings with elevated spans over an area to support a built structure.

Pile A vertical support *footing* sunken or rammed into the ground for a built structure.

Pillar roots See *Column roots*.

Pine cone See *Cone*.

Pinocchio nose rib See *Sharp-edged rib*.

Pioneer roots *Lateral roots* of some trees that grow rapidly for considerable distances beyond the crown with little branching or taper maximising the potential access to a greater soil volume for water and nutrient uptake, e.g. *Magnolia grandiflora* (Thomas 2000, p. 77).

Pioneer tree *Locally indigenous* tree that initially develops in an *ecological community*, from which it derived, after that community has been disturbed (Robinson 2004, pp. xxix and 145). As opposed to *non-locally indigenous* or *exotic* species that readily colonise disturbed locations often as *nuisance* or *weed species*.

Pith *Parenchyma* tissue usually located central to a branch or stem, rarely evident in roots of trees, primarily for storage of carbohydrates.

Plagiotropism A directional growth movement of a plant or plant part at an angle away from the vertical in response to stimulus.

Planned maintenance Tree maintenance undertaken as part of a programmed and budgeted schedule of works. See also *Unplanned maintenance* and *Tree management*.

Planning instrument A list of legislative or *policy* obligations used for planning decision making.

Planning provision A clause in *legislation* prescribing that a condition must be met.

Planting strip The area of the *road reserve* dedicated to the cultivation of trees, grass or other plants and usually located in a narrow section of land located immediately behind and parallel to the kerb.

Plant pathologist Individual specialising in the scientific study of plant diseases and abnormalities.

Plant pathology See *Phytopathology.*

Plan view Plan representing a design when viewed from above.

Pleaching The weaving, intertwining or grafting of branches into various forms which are then maintained by continual pruning, e.g. an arch formed as an *arbour* by trees planted on both sides of a path (Australian Standard 2007, p. 7).

Pneumatophore An erect modified root structure protruding some distance above soil level being produced in great numbers by some trees such as mangroves in tidal areas or saturated soils, e.g. *Avicennia marina* and *Taxodium distichum* respectively. These structures are perforated by many *lenticel*s that promote *gaseous exchange* between the roots and the atmosphere.

Pocket crotch A generally concave *crotch* where moisture and *leaf litter* accumulate and subsequently *humus* often forms supporting *fine roots* as *adventitious roots* acting as a *potential habitat tree* for some ferns and orchids. See also *Phytotelmata.*

Pointy nose rib See *Sharp-edged rib.*

Policy A planning and management guideline on a specific issue/s to assist in decision making, e.g. tree management policy.

Pollard head The enlarged stem area of *wound wood* formed below a succession of *pruning wounds* especially by *pollarding*, containing *latent* or *adventitious buds* that arise in response to the stimulus of each *pollarding* episode.

Pollarding A pruning technique to establish branches that terminate with a *pollard head*, from which arise multiple vigorous shoots (Australian Standard 2007, p. 7).

Pollen cone Male flower structure of some gymnosperms.

Poor condition Tree is of good habit or *misshapen*, a form that may be severely restricted for space and light, exhibits symptoms of advanced and *irreversible decline* such as fungal, or bacterial infestation, major dieback in the branch and *foliage crown*, *structural deterioration* from insect damage, e.g. termite infestation, or storm damage or lightning strike, ring barking from borer activity in the trunk, root damage or instability of the tree, or damage from physical wounding impacts or abrasion, or from altered local environmental conditions and has been unable to adapt to such changes and may decline further to death regardless of remedial works or other modifications to the local *environment* that would normally be sufficient to provide for its basic survival if in *good* to *fair* condition. Deterioration physically, often characterised by a gradual and continuous reduction in vigour but may be independent of a change in vigour, but characterised by a proportionate increase in susceptibility to, and *predation* by pests and diseases against which the tree cannot be sustained. Such conditions may also be evident in trees of advanced senescence due to normal phenological processes, without modifications to the growing environment or physical damage having been inflicted upon the tree. This may be independent from, or contributed to by vigour. See also *Condition*, *Good condition* and *Fair condition*.

Poor form Tree of *atypical* crown shape and habit with proportions not representative of the species considering constraints and appears to have been adversely influenced in its development by environmental factors *in situ* such as *soil water* availability, prevailing wind, cultural practices such as lopping and competition for space and light; causing it to be *misshapen* or disfigured by disease or vandalism. See also *Good form*.

Poor vigour See *Low vigour*.

Population A group of organisms, all of the same species, occupying a particular area.

Pore The spaces between *soil* particles, within or adjacent to a *ped*. Pore spaces can be categorised as *Micropores*, *Mesopores* and *Macropores*.

Pore space See *Pore*.

Porous paving Structures used as pavement that allow water to pass through, e.g. bonded aggregates, contained aggregate, stabilised turf, open jointed block systems and soft porous surfacing (Thompson & Sorvig 2008).

Potential habitat tree Any tree that develops a niche suitable to provide support for the life processes of a plant or animal, e.g. hollows in the trunk or branches, suitable for nesting birds, arboreal mammals and marsupials, e.g. squirrels, bats or possums, or support of the growth of epiphytic plants, e.g. orchids and ferns. See also *Habitat tree.*

Pre-cast concrete lintel A proprietary manufactured concrete lintel. Such lintels are often produced in a range of lengths. See also *Lintel.*

Pre-cutting Undercuts, side cuts and/or scarf cuts made away from the *branch collar* prior to the *final cut* to reduce the risk of a branch splitting or tearing (Australian Standard 2007, p. 7).

Predation Temporary or prolonged attack by pests or diseases resulting in physical damage and disruption to life processes. See also *Resistance* and *Vigour.*

Preliminary drill/s Initial cross-sectional drilling undertaken by a *Resistograph* to determine the maximum cross-sectional area of *decay* within a trunk or branch. Generally carried out when no external signs of internal decay exist, such as bulging or cavities which may be evident using *Visual tree assessment.* Undertaken sequentially along the long *axis* of the trunk or branch whereby the residual wall thicknesses are observed for an increase or decrease which will determine the location for further cross-sectional drilling if required. See also *Reference drill.*

Prelodgement A preliminary assessment by the *consent authority* at the request of the *applicant* prior to formal submission of a *development application* to verify compliance with planning requirements.

Premature aging Apparent hastened aging and deterioration of a tree where it has been subject to conditions or practices adverse to expected normal growth, resulting in a *spiral of decline.* The following are examples of processes that may start such cycles:

- Top lopping of a mature tree

- In a new car park, the excavation of soil severing the roots of a tree close to its trunk and then sealing the soil surface with asphalt or concrete up to the trunk

- Open trenching alongside a street tree severing all roots in the trench, then top lopping it for power line clearance, and then extensive damage to bark by abrasion by trucks and excavation equipment as tree is adjacent to a construction site

- Root damage from *soil compaction* to substantial areas of the root plate.

Premature senescence See *Premature aging.*

Pressler increment borer A hollow auger-like instrument with an extractor that is used to remove a thin increment core (usually 3 mm diameter) from a *stem* for examination, named after its inventor in 1867.

Prestressing The existing and increasing loading of trunk and branches as the mass of a tree gradually increases through *growth increments* where a trunk or branch is compressed on the inside and under tension on the outside (Ennos 2001, pp. 37–38).

Previous season's growth See *Last season's growth.*

Primary growth Elongation of root growth by rapid cell division at the *root tip* where little radial expansion occurs (MacLeod & Cram 1996, p. 1). Extensive branching of these roots forms the *adventitious root mass* in palms.

Primary phloem Tissues conducting food produced by growth activities originating in apical meristems.

Primary roots The initial root system developed from the *radicle* being the lateral *first order roots* at the *root crown* and taproot. In palms the initial development of adventitious roots.

Primary root zone (PRZ) 1. Minimum root mass and soil volume essential for the basic survival of a tree, enabling it to be sustained or retained in *good condition*, without alteration to its *typical* physical characteristics or *stability*. 2. A method that considers a minimum radial distance from the trunk that excavation as *cut* and *fill* and construction are permissible to enable a tree to be satisfactorily retained. For this method satisfactory setbacks are usually considered as 10 times (10×) *DBH* with a minimum setback of 2 m with the possibility of cut, fill or trenching on one side only (tangentially) to five times (5×) *DBH* as an incursion into the area after further *structural root* examination. However, this does not consider *age*, *condition* and *vigour*.

Primary thickening In a stem this is the relatively narrow zone of *meristematic* cells producing radial derivatives, usually *parenchyma* towards the outside and both *parenchyma* and discrete vascular bundles towards the inside. The responsibility of this primary thickening meristem is initial stem thickening, adventitious root production, and formation of linkages between roots, stem and leaf vasculature (Rudall 1992, pp. 26–28).

Primary xylem Water-conducting tissue where differentiation has not yet occurred to form *vascular cambium* and may be evident by growth activities originating in *apical meristems*.

Private open space Open space in private ownership that is usually inaccessible to the public. See also *Public open space*.

Procambium The part of an *apical meristem* that gives rise to primary vascular tissues.

Profile See *Soil profile*.

Progeny The descendants of a plant, e.g. the seeds produced by a tree.

Progressively leaning A tree where the degree of *leaning* appears to be increasing over time.

Prolonged senescence A phenomenon in an *over-mature* tree or tree with *structural deterioration* in its *condition* and often *vigour* as *abnormal vigour* as a result of modifications to the tree or the growing environment essential for its survival where it is sustained beyond the *typical* extent of its life cycle, or prevented from failing in full or part from *structural deterioration* by a beneficial artificial modification to its growing environment either by deliberate or incidental intervention, e.g. water from a leaking tap, water and nutrients from a leaking sewer pipe creating a *hydroponic* environment, or by physically propping up a tree with *structural deterioration* as with a *veteran tree*, or by it *leaning* or growing against another tree or structure for support.

Propagation The methods used to multiply specimens of a plant from parent stock, either sexually, e.g. from seed, or asexually (vegetatively), e.g. from tissue culture, cutting or a *graft*.

Prop roots Adventitious roots or *column roots* that developed from the lower nodes of the stem or trunk in some trees serving to provide additional support, e.g. Monocotyledons *Pandanus* ssp. and some palms, e.g. *Cryosophila warscewiczii*, where the base of the trunk may wither leaving the aerial parts supported by these roots.

Protection wood *Sapwood* altered to a higher chemical state of protection as initiated by processes of aging, death of a branch, or wounding and infection (Shigo 1986, p. 87).

Protective tissues See *Bark*.

Proteoid roots A modification to root morphology in the *Proteaceae* and *Restionaceae* families in dry, nutrient poor, heath land soils, often sandy, forming a cluster of roots in the superficial horizons of a *soil profile* (humic layer) utilising actively decomposing litter to most likely enhance the capture of Phosphorous, and in its inorganic form (Smith & Read 1997, pp. 419–422).

Provenance A particular geographic site or region from where something originates (Boland *et al.* 2006, p 696).

Proximal A section of any tree part, closest to its point of attachment. See also *Distal*.

Pruning Removal of any branch or root, dead or alive, by severance across the stem, back to the intersection of another live stem to a swollen area at the intersection called a *branch collar* where such a structure exists, with a *final cut* at the outer edge of the collar leaving no stub, or to undamaged woody tissue for *roots*. Also the severing of any part of a tree so as to cause a reduction of the air space occupied by the branches and foliage in the *crown* or roots in the *root plate*. Examples of pruning are deadwooding, crown lifting, formative pruning, reduction pruning, selective pruning, crown thinning, and remedial and *restorative pruning* (Australian Standards 2007, p. 6). Pruning should conform to recognised standards, e.g. Australian Standard® AS 4373 'Pruning of Amenity Trees'. The following are not recognised as pruning: *lop, lopping, top, topping, top lopping.*

Pruning wound A wound created by the act of *pruning.*

Pseudo-street tree A tree not growing in the *road reserve* but in adjoining private or public land and growing so that its crown develops and extends to contribute substantially to the streetscape as a false street tree (Draper 1997, pp. 6–10). See also *Street tree.*

Public consultation A process to gauge public opinion and interest in matters of community interest before decisions are made, e.g. prior to mass tree planting or removal in a park or street.

Public open space Open space in public ownership that is usually accessible to the public, e.g. parks, bushland reserves, national parks, cemeteries. See also *Private open space.*

Radar 1. Acronym – radio detection and ranging. Radio waves are bounced off an object, and the time at which the echo is received indicates its distance (Howard University 2006). 2. In trees, utilising electromagnetic characteristics rather than density to produce echographs from a pulse generator and an antenna placed on opposite sides of a *stem*, reliably detecting cavities but with limited sensitivity to changes in wood density (Nicolotti & Miglietta 1998, pp. 297–302).

Radial The directions from the centre of a circular object to the outer edge, such as in a branch, trunk, root, root plate or crown.

Radial core *Increment core* taken in a line orientated towards the centre of a *stem* (Weber & Mattheck 2003, p. 44).

Radial crack Narrow splitting usually longitudinally along a *stem*, internal in origin, and may continue for some distance (Mattheck & Breloer 1994, pp. 104–105).

Radial mulching See *Radial trenching.*

Radial trenching A series of excavated trenches near to the trunk, usually between *first order roots*, allowing space for the introduction of improved *growing media.* This process is designed to stimulate new root growth through reduced compaction, improved aeration and removal of contaminated soil (Roberts *et al.* 2006, p. 108). See also *Vertical mulching.*

Radical See *Radicle.*

Radicle The seed root of angiosperms and gymnosperms that give rise to the primary root system. The radicle may persist forming a taproot or be replaced by lateral or adventitious roots.

Ragged per cent See *Crown integrity.*

Rainfall interception See *Interception.*

Ram's horning *Wound wood* that becomes curled inward and can wrap around itself as it crosses a void such as a *cavity* and may succumb to cracking with those wounds susceptible to further infestation by *decay* pathogens.

Ray cells See *Vascular ray.*

Rays See *Vascular ray.*

Reaction wood A negative geotropic response in some *secondary xylem* to counter a lean or predominant mechanical force, formed as *tension wood* in dicotyledonous *angiosperms* and as *compression wood* in *gymnosperms* (Figure 24). See also *Adaptive wood.*

Reaction zone A protective boundary zone usually dark in colour within the wood of a living tree, between fully functional sapwood and dysfunctional or decaying wood (Shigo 1986, p. 91; Lonsdale 1999, p. 319).

Reactive maintenance See *Unplanned maintenance.*

Reactive wood See *Reaction wood.*

Reactive zone See *Reaction zone.*

Reduced level (RL) A point of reference on the Earth's surface taken from a datum point, e.g. *Australian height datum* (AHD).

Reduction pruning Branches specifically pruned to reduce crown height or *crown spread* by *pruning* to reduce the length of a branch with a *final cut* at

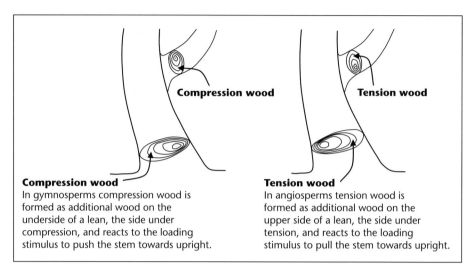

Compression wood
In gymnosperms compression wood is formed as additional wood on the underside of a lean, the side under compression, and reacts to the loading stimulus to push the stem towards upright.

Tension wood
In angiosperms tension wood is formed as additional wood on the upper side of a lean, the side under tension, and reacts to the loading stimulus to pull the stem towards upright.

Figure 24 Reaction wood – compression wood and tension wood.

a branch union inside the crown (Australian Standard 2007, p. 8). Here the retained branch should be greater than one-third ($>\frac{1}{3}$) of the diameter of the removed branch. Note: *lopping* is not crown reduction. See also *Lopping.*

Reference drill An initial drill undertaken in *sound wood* of a tree to be tested for decay with a *Resistograph* to establish the typical *sound wood* reading for that tree prior to testing (Blank 2006).

Regional environmental plan (REP) In Australia, a *statutory plan* prepared by the state or federal government which addresses matters of regional significance.

Remedial action See *Mitigation.*

Remedial pruning Pruning to repair previously poorly undertaken works or to assist in re-establishing the *crown form* and shape of a tree that has been damaged, or exhibits *dieback*. Pruning may require a *final cut* beyond the branch collar to stimulate epicormic shoots from which the new crown structure is developed by *reduction pruning* or *crown thinning* (Australian Standard 2007, pp. 14, 15).

Remedial surgery See *Remedial pruning.*

Remnant See *Remnant vegetation.*

Remnant vegetation A plant or plants of any taxa and their progeny as part of the floristics of the recognised endemic *ecological community* remaining in a given location (e.g. seeds in seed bank, trees) after alteration of the site or its modification or fragmentation by activities on that land or on adjacent land, e.g. trees and *bushland* isolated after land clearing for rural or urban development.

Removal/remove 1. Processes carried out for dismantling and disposal of a live or dead tree in full or part *in situ,* e.g. by the use of a chainsaw, or by dislodging it from its growing location with earth moving equipment. This may also include the poisoning of the stump and/or roots and/or taking away, or grinding or burning out of its remains to prevent regrowth. 2. Displacement of a tree from its growing location in order to relocate and sustain it by *transplanting.*

Removed No longer present, or tree not able to be located or having been cut down and retained on a site, or having been taken away from a site prior to site inspection.

Resin The extractive substance produced by most gymnosperms contained within long resin ducts. Resin plays a role in herbivore defence by inhibiting the activity of wood boring insects.

Resistance The ability of a tree to withstand particular adverse conditions or attack by a specific pest or pathogen. See also *Vigour* and *Predation.*

Resistograph® See *Penetrometer.*

Resonant frequency The natural frequency of swaying in a tree (Lonsdale 1999, p. 319) when exposed to mean wind speed as gusts and lulls over a period of time (Australian Bureau of Meteorology 2008).

Respiratory roots See *Pneumatophore.*

Respondent Individual or party defending their position in a matter of disputation brought before a court for resolution by an *applicant.*

Restorative pruning See *Remedial pruning.*

Rest point The position of a tree when stationary, e.g. not moved by wind.

Rhizosphere Root and soil interface surrounding the surfaces of *fine roots* influenced by interactions between roots and soil micro-organisms, extending >1–10 mm from the root surface (Craul 1992, pp. 52, 128).

Rhytidome Rough outer bark of many types of tree formed as secondary *periderm* (Lonsdale 1999, p. 320).

Rib *Adaptive wood* that may form over a *crack, included bark* or *enclosed bark* and may be a *sharp-edged rib* as an elongated protuberance where a crack continues to develop or a *round-edged rib* where a broad convex swelling is formed over the crack by the addition of each new *growth increment* and the cracking is slowed or prevented from developing further (Mattheck & Breloer 1994, p. 57). Some rib-like growths may not be related to *cracks* or *included bark* having formed by older enlarged *aerial roots*, e.g. *Melaleuca quinquenervia.* (See Figure 25 on p. 127.)

Right of way Land which provides for the location of utilities above or below ground, or access through part of one property to another.

Ringbark An encircling *wound* usually inflicted around the circumference of the trunk (except monocotyledons) to a sufficient depth to permanently disrupt the *vascular cambium*, usually resulting in the death of the tree.

Ringbarked Any process where an encircling *wound* has been inflicted around the circumference of a stem or root at a sufficient depth to effect permanent disruption of the *vascular cambium*.

Ringbarking See *Ringbark*.

Ring porous One of the two main types of wood structure in broadleaf trees in which each annual increment includes two distinct bands when seen in cross-section, these consist of *early season's wood* with wide *vessels*, and of *late season's wood* with narrow vessels, e.g. Oak, Ash, False Acacia. See also *Diffuse porous*.

Ring-scars See *Annular rings*.

Ring shake Internal separation of wood along *growth rings*, usually where a *barrier zone* has formed creating an area of structural weakness (Shigo 1989a, p. 152).

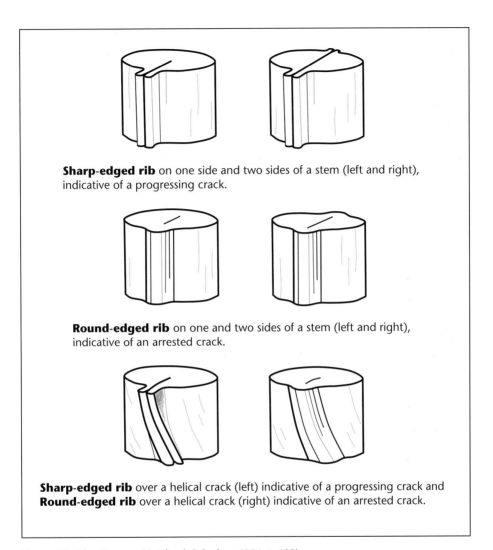

Sharp-edged rib on one side and two sides of a stem (left and right), indicative of a progressing crack.

Round-edged rib on one and two sides of a stem (left and right), indicative of an arrested crack.

Sharp-edged rib over a helical crack (left) indicative of a progressing crack and **Round-edged rib** over a helical crack (right) indicative of an arrested crack.

Figure 25 Ribs. (Source: Mattheck & Breloer 1994, p. 182)

Ring splits See *Ring shake.*

Ring swelling Occurs where *adaptive wood* forms around a stem to bridge over and conceal decay, to maintain *structural integrity* in accordance with the *axiom of uniform stress*, evident as an exaggerated localised swollen band (Mattheck & Breloer 1994, pp. 99–101).

Ripewood The older central wood of those tree species in which sapwood gradually ages without being converted to *heartwood* (Lonsdale 1999, p. 320).

Risk The random or potentially foreseeable possibility of an *episode* causing harm or damage.

Road Route trafficable by motor vehicles; in law, the public right-of-way between boundaries of adjoining property (Australian Standards 2002).

Road reserve The land between the edge of the street or roadway, determined by the presence of a kerb and gutter where formed, and the adjoining property boundary. This land is usually set aside for road widening or for the location of aerial or subterranean utility services and the planting of street trees.

Road shoulder The outer edge of a roadway usually chamfered to allow for drainage of the road surface. Sometimes trees may be planted in this area.

Roadway See *Road*.

Rock floater A large section of rock detached from the *bedrock* usually buried or partially buried.

Rock outcrop A *rock floater* or *bedrock* evident at or protruding from the ground surface.

Root ball 1. Within the *root plate*, the volume of roots and soil (in part or full) retained when a tree is excavated for transplanting. 2. The volume of roots and soil of a containerised plant.

Root barrier A physical or chemical obstruction located in the ground to prevent or divert the spread of roots, usually to protect nearby infrastructure. Root barrier types are usually *trapping, deflecting or inhibiting* (Coder 1998, p. 63; Roberts *et al.* 2006, p. 355).

Root bud *Adventitious buds* formed from *wound wood* in damaged roots that may remain dormant or develop to form clone-like trees from *sucker shoots*, e.g. *Robinia pseudoacacia*.

Root cap See *Calyptra*.

Root collar The ring of growth sometimes formed around the *proximal* end of a root by the cells of the trunk at the *root crown* or a root from which a lower order root arises.

Root crown Roots arising at the base of a trunk.

Root crown inspection Comprehensive examination of the lower trunk, root crown, the area below and surrounding to determine structural integrity, *stability* and the presence of pathogens.

Root deflecting See *Root barrier* and *Deflecting*.

Root flare See *Basal flare*.

Root graft The *graft* formed between woody roots from the same tree or different trees nearby, usually of the same species, where they grow against one another and merge when sufficient pressure is applied. Vascular functions and structural loading may be shared allowing the movement of plant growth regulators (hormones) through *phloem* (Thomas 2000, pp. 94 and 97) and assimilates and water through *xylem*. While such a process may be beneficial it may also be detrimental due to the risk of *backflash*. This process may also be deliberate as a horticultural process, or incidental, e.g. *Ficus* spp. See also *Anastomosis*. This also occurs with parasitic or hemi-parasitic species of tree via *haustorium,* e.g. *Santalum, Nuytsia* and *Exocarpos*. See also *Grafted root zone*.

Root grafting See *Root graft*.

Root hairs Hair-like structures formed from single epidermal cells behind the *root tip* that grow to be in direct contact with the soil. This increases the surface area of roots and is the primary site on *fine roots* for the uptake of *soil water* and nutrients in solution. Root hairs are usually short lived and are replaced as the root tip elongates. Roots with a *mycorrhizal association* may produce fewer root hairs. See also *Fine roots*.

Root mapping The exploratory process of recording the location of roots usually in reference to a *datum point* where depth, root diameter, root orientation and distance from trunk or structures are measured. It may be *invasive root mapping* or *non-invasive root mapping*.

Root mass See *Root plate*.

Root nodule A small swelling on a root resulting from invasion by symbiotic nitrogen-fixing micro-organisms such as bacteria.

Root plate The entire root system of a tree generally occupying the top 300–600 mm of soil including roots at or above ground and may extend laterally for distances exceeding twice the height of the tree (Perry 1982, pp. 197–221). Development and extent is dependent on water availability, soil type, *soil depth* and the physical characteristics of the surrounding landscape.

Root pressure The pressure developed by living cells in a root forcing water up the xylem, driven by *transpiration* from *photosynthesis*.

Root protection area (RPA) See *Root protection zone (RPZ)*.

Root protection zone (RPZ) A specified area on and below the ground and at a given distance from the trunk set aside for the protection of tree roots to provide for the *viability* and *stability* of a tree to be retained where it is potentially subject to disturbance by development. Establishment of this area may include root mapping, root pruning and installation of root barriers or other remedial works at the edge of the RPZ to prevent conflict between roots and works.

Root rot *Decay* of *structural roots* or disease killing fine roots. See also *Fungus*.

Roots The part of a tree either woody or fibrous, branched, growing mostly under the ground providing mechanical support for anchorage, storage of energy reserves and the uptake of water and elements in solution. The roots on most of the tree species of the world grow within the upper 1 m of the soil. Many tree roots are very shallow and some roots in tropical areas grow for long distances along the soil surface. Roots on most trees usually spread far beyond the *dripline* of their *crown* and may extend for distances in excess of two to three times the mature height of a tree (Perry 1982, pp. 197–221). Roots may be woody as with gymnosperms (conifers) and dicotyledonous angiosperms (flowering plants) or *adventitious roots* with monocotyledonous angiosperms (palms). Roots develop by *primary growth* in palms, and by *primary growth* and *secondary growth* in other trees (MacLeod & Cram 1996, p. 1).

Root–shoot ratio Relative proportion of mass between the roots and crown (or shoots) on a weight basis (Harris *et al.* 2004, p. 532).

Rootstock See *Understock*.

Root tip The apical few millimetres or less at the end of *fine roots* that elongate to grow through the soil.

Root zone See *Root plate*.

Round-edged rib *Adaptive wood* formed over a *crack,* e.g. *hazard beam crack*, or *branch bark inclusion* or *enclosed bark* where the cracking is arrested or slowed by the addition of each new *growth increment* forming a broad convex swelling. Such ribs may be straight or *helical* (Mattheck & Breloer 1994, pp. 57, 104–105 and 182). See also *Rib* and *Sharp-edged rib,* and Figure 25 on p. 127.

Round-edged rib and Sharp-edged rib combination These may occur as a combination of both *round-edged rib* and *sharp-edged rib* when the crotch of the branch bark union is included on one side of the fork and has a weak attachment on the other. This is generally a weak branch union. This may result in *collapse and hinge branch failure* (Mattheck & Breloer, 1994, pp. 64 and 65). See also *Rib, Round-edged rib* and *Sharp-edged rib.*

Rounding over See *Lopping*.

S

Sabre tree A tree generally growing on a slope affected by static or dynamic loading by seasonal accumulation of snow on the ground pushing against the trunk. This causes it to lean down slope where it develops a progressive curvature or *butt sweep* in the trunk as negative gravitropism stimulates *self-correcting* growth to return the crown to upright.

Safe useful life expectancy (SULE) A now redundant system to determine the time a tree can be expected to be usefully retained. SULE was based on a system designed to classify trees into a number of defined categories so that information regarding tree retention could be concisely communicated in a non-technical manner for application in a planning context. SULE categories were easily verifiable by experienced personnel without great disparity. A tree's SULE category was the life expectancy of the tree modified first by its age, health, condition, safety and location (to give safe life expectancy). Then by economics (i.e. cost of maintenance; retaining trees at an excessive management cost is not normally acceptable), effects on better trees, and sustained amenity (i.e. establishing a range of age classes in a local population). SULE assessments were not static and able to be modified as dictated by changes in tree health and environment. Trees with short SULE may have presently made a contribution to the landscape but their value to the local amenity would decrease rapidly towards the end of that period, prior to their being removed for safety or aesthetic reasons (Barrell 1993 and 1995). See also *TreeAZ* and *Sustainable retention index value* (SRIV).

Sail area Total surface area of the *crown* exposed to wind.

Sap A fluid consisting of mineral salts and *sugars* dissolved in water that is formed in *xylem* and *phloem vessels*.

Sapling A *young* tree, early in its development with small dimensions.

Saprophyte An organism acquiring organic matter in solution from dead and decaying tissues of other organisms.

Saprophytic See *Saprophyte*.

Saprot Fungal *decay* of sapwood.

Sapwood The living cells in the outer, usually light-coloured, water and nutrient conducting wood forming region of *secondary xylem*.

Scaffold branches See *Structural branches*.

Scaffold limbs See *Structural branches*.

Scar tree A tree containing a wound of cultural or scientific interest, inflicted initially for a specific purpose, e.g. by indigenous people to extract implements or carved as a marker or with a pattern for ceremonial purposes, or as a marker and *blaze* by a surveyor or explorer, or from an accidental *wound* that has not *occluded*.

Scion A crown section of a tree selected for propagation by grafting for its specific horticultural characteristics, e.g. *crown form*, fruit, flower or foliage colour, as a *graft* onto a vigorous *understock*.

Sclerenchyma Principal supporting cells in stems that have ceased elongation where the cell walls have become lignified and thickened.

Screw pile A tubular steel *pile* with a helix edge towards its base that is rotated vertically into the ground bedding it into the substrate with the remaining section above ground incorporated into the footings for a built structure.

Often used in confined spaces and in unconsolidated soil such as sand. It produces no spoil and may minimise root disturbance.

Scribble wound Superficial scribble-like wounding from the innocuous larvae of *Ogmograptis scribula* – Scribble Moth, burrowing in irregular patterns between bark layers and revealed as bark *decorticates*. Usually evident on smooth-barked species of Eucalyptus with some known as Scribbly Gums being *Eucalyptus racemosa, E. haemastoma* and *E. rossii* and affecting other trees *E. pauciflora* and *E. mannifera subsp. maculosa* (Jones & Elliot 1986, pp. 112–113; Brooker & Kleinig 1999, pp. 145, 333–335, 342).

Scribing The removal of injured bark and wood from around *wound margins* (Shigo 1986, p. 98) cleaning the edges to promote *occlusion*.

Seam See *Crotch seam.*

Secondary crown Evident often during *senescence* or after *dieback* as a result of an *episode* of severe *stress* where *epicormic shoots* are produced *proximally* from first and lower order branches usually in the *mid crown* to *lower crown*. Presumably to save energy by the *translocation* of reserved carbohydrates and photosynthates over shorter distances developing a crown of reduced dimensions with the appearance of being retracted within the original crown. Remnants of the original crown usually remain protruding above the new crown but continue to *decline* as the *crown cover* and density of the secondary crown becomes dominant, e.g. *Araucaria bidwillii*, and *Eucalyptus nicholii*. See also *Crown regeneration.*

Secondary growth In gymnosperms and dicotyledonous angiosperms the growth in diameter of roots after *primary growth* in mature parts of a root system that undergo *secondary thickening* (MacLeod & Cram 1996, p. 1).

Secondary phloem Tissue formed by the *vascular cambium* for conducting food.

Secondary thickening After *primary thickening* (excluding palms) the growth of the *cambial zone* forming rings around the stem. Here *xylem* forms on the inner side to provide structural support to the stem developing as sapwood for the movement of water and nutrients to the *foliage crown*, and *phloem* on the outer side for trunk protection and movement of *sugars* to the root system.

Secondary xylem Water conducting tissue in trees formed by the *vascular cambium*.

Seed root system See *Radicle*.

Selective pruning The removal of identified or specified branches (Australian Standard 2007, p. 8).

Selective removal The removal of specific tree/s or particular classes of trees, in a planned manner, e.g. declining, hazardous, weed species, dead, nurse plantings, thinning of a stand. See also *Alternation of generation*.

Self-corrected lean See *Self-correcting*.

Self-correcting *Atypical* stem growth subsequently influenced and modified by *tropisms*, e.g. *gravitropism* and *phototropism*, where *reaction wood* attempts to return it to a more *typical* habit or *form*, e.g. a trunk with a *butt sweep* where it is returning to upright.

Self-graft See *Grafted branches*.

Self-optimising See *Axiom of uniform stress*.

Self-sown A plant established by itself either vegetatively or from a seed, without human intervention initially in its propagation or cultivation. Such plants are often vigorous and may have the potential to develop as weeds. This may be the successful progeny of a planted tree.

Semi-deciduous A tree that sheds all or most of its leaves in one season and enters into a short *dormant* period, usually sprouting new buds soon after the old leaves have been abscised.

Semi-mature An intergrade between the age classes of *young* and *mature*.

Semi-parasite Organism that is only partly parasitic on its host, e.g. *Mistletoe*. Such an organism may acquire some of its food from being parasitic on the stems of its host (e.g. mistletoe), or on the roots of other trees, shrubs or grasses and also produces its own food through photosynthesis or may require a host to become established, e.g. *Exocarpos cupressiformis* and *Santalum acuminatum*. Severe infestations may weaken and eventually kill the host.

Senescence See *Senescent.*

Senescent Tree of advanced old age, or over-mature leading towards death.

Severely crooked Stem section growing at a lean 60–90° from upright, *atypical* from its natural habit or *crown form*. See also *Crooked, Slightly crooked* and *Moderately crooked.*

Severely leaning A leaning tree where the trunk is growing at an angle within 30–45° from upright. See also *Leaning, Slightly leaning, Moderately leaning* and *Critically leaning.*

Shade tree A tree planted specifically for the shadow cast by its *crown* especially in summer, often utilising a deciduous tree conversely to provide *solar access* in winter while *dormant,* and may be synonymous with *amenity tree/s* in the USA.

Shag A collective term for the persistent dead fronds that hang from some palm taxa (Jones 1996, p. 270).

Shakes Wood separation in a tree along a circumferential plane or a radial plane (Shigo 1986, p. 99). See also *Ring shake.*

Shallow rooters Trees with a shallow *root plate* morphology that develops buttresses for *stability* with *sinker roots* for *anchorage* towards the outer edge of the *root plate.* Such trees may be subject to *windthrow* where the soil cracks, roots slide and the soil plate tears out and tips with minimal shearing (Mattheck & Breloer 1994, pp. 72–73).

Shallow soil Soil to a depth of 500 mm or less (Craul 1992, p. 32). See also *Deep soil.*

Sharp-edged rib *Adaptive wood* forming over a *crack,* e.g. *hazard beam crack,* or *branch bark inclusion* or *enclosed bark,* where the crack continues to form promoting an elongated swelling. Such ribs may be straight or *helical* (Mattheck & Breloer 1994, pp. 57, 104–105 and 182). See also *Rib* and *Round-edged rib,* and Figure 25 on p. 127.

Shear failure A plane of weakness within a structure, e.g. a trunk, where sections slide against one another. Sometimes evident in a tree that is straight but *leaning* (Mattheck & Breloer 1994, p. 195).

Shearing See *Shear failure.*

Shear stress Loading force that promotes *shear failure.*

Shedding See *Abscission.*

Shell buckling Occurs as *structural failure* and potential *collapse* of a thin walled *hollow* stem buckling initially, then splintering into several fracture-planks that each subsequently fail from kinking (Mattheck & Breloer, 1994, pp. 31–33).

Shigometer® See *Electrical conductivity meter.*

Shoot A juvenile branch bearing leaves.

Shore To prop something up to prevent its failure or collapse, e.g. utilised with excavation in sand or unstable soil.

Shoring See *Shore.*

Short term A period of time less than <1–15 years. See also *Periods of time, Immediate, Medium term* and *Long term.*

Shredded wood Wood separated into narrow strips up to 2 mm wide by insects such as *Aesiotes leucurus* – Cypress Bark Weevil, as their larvae feed beneath the bark in the *phloem* area of the *vascular cambium* in some *Cupressus* spp. and some other conifers.

Shrub A woody *perennial* plant often long-lived, usually *acaulescent*, branches *deliquescent*, and generally less than 5 m high (Brock 1993, p. 346). See also *Tree.*

Sibert® See *Penetrometer.*

Sidewalk Synonymous with *road reserve* and footpath in the USA.

Sieve tube A food conducting cell.

Significant Important, weighty or more than ordinary.

Significant tree A tree considered important, weighty or more than ordinary. Example: due to prominence of location, or *in situ*, or contribution as a component of the overall landscape for *amenity* or aesthetic qualities, or *curtilage* to structures, or importance due to uniqueness of taxa for species, subspecies, variety, *crown form*, or as an historical or cultural planting, or for age, or substantial dimensions, or habit, or as *remnant vegetation*, or habitat potential, or a rare or threatened species, or uncommon in cultivation, or of Aboriginal cultural importance, or is a commemorative planting.

Significant tree register A recorded list or data base of *significant tree* specimens or stands which may be compiled at a municipal, state or national level, usually protected by planning or *heritage* laws.

Silviculture The growth of trees for purposes other than for *amenity* and utility, usually commercially, e.g. for timber, paper or oil production.

Sinker roots Roots growing vertically downward usually from a *lateral root* at short distances from the trunk, often in the *zone of rapid taper*, in search of water, providing additional support and *anchorage* to the *root plate*.

Site See *The site*.

Site arborist A *consulting arboriculturist* or *arborist* appointed to a development site, often as a requirement of development consent, specifically for the management of trees throughout all phases of a *development* including design, demolition, preparation, construction and completion including ongoing monitoring and reporting throughout the different phases of the development. This may also include long-term management of the trees for extended periods following completion of works including monitoring, reporting and arrangement of remedial works.

Site hearing Part of a court hearing spent visiting and examining a property that is the subject of dispute in a legal matter regarding environmental planning.

Size of tree The relative proportions of a tree based on a combination of its dimensions at maturity, *in situ*: height, *crown spread*, *trunk* and structural branch diameters or a combination of these. The size of a tree can be categorised as *small tree*, *medium tree* and *large tree*.

Slightly crooked Stem section growing at a lean 0–30° from upright, *atypical* from its natural habit or *crown form*. See also *Crooked*, *Moderately crooked* and *Severely crooked*.

Slightly leaning A leaning tree where the trunk is growing at an angle within 0–15° from upright. See also *Leaning, Moderately leaning, Severely leaning* and *Critically leaning*.

Small deadwood A dead branch up to 10 mm diameter and usually <2 m long, generally considered of low *risk* potential. See also *Deadwood* and *Large deadwood*.

Small tree A tree with a height <10 m or *crown spread* <10 m at maturity, *in situ*. See also *Size of tree, Large tree* and *Medium tree*.

Snag A standing *dead* tree, or a broken branch stub.

Snub nose rib See *Round-edged rib*.

Soft rot Decomposition by fungi of only a portion of the cell wall components of wood, and here largely *cellulose* or *hemi-cellulose* may be attacked in localised pockets in the secondary cell wall and occurs commonly in wood saturated by water or in direct contact with soil. This is characterised by surface softness and formation of shallow cross-checking in the wood upon drying. This rot is caused by fungi from the subdivisions *Ascomycotina* and *Fungi imperfecti* (Manion 1991, pp. 227–228).

Soft soil Area of soil volume set aside for landscape planting.

Softwood See *Gymnosperms*.

Soil Aggregate material formed over time and accumulated usually in layers on top of the Earth's land surface, formed in a natural *environment* from weathered rock and consisting of minerals, organic material, living organisms, air and water, supporting the growth of terrestrial plants. Variation on soil formation depends on climate, parent rock and its mineral properties and their resistance to weathering both chemical and mechanical, topography, rainfall and latitude. The particle sizes of the rock content in soil are described generally from smallest to largest as clay, silt, sand and gravel.

Soil ameliorant See *Soil amendment*.

Soil amendment Material mixed with soil to adjust its physical or chemical properties.

Soil analysis Assessment of a soil sample for *bulk density, pH, soil fertility* available phosphorus, contaminants and exchangeable cations (Craul 1992, p. 350; Handreck & Black 2002, p. 437).

Soil compaction Pressing and squashing of soil that removes macropore spaces, eliminating its water and air holding capacity resulting in an increase in *bulk density* and damage to *structure*.

Soil core sample Extracted cylindrical specimen of soil, usually taken vertically to a given depth from the surface, for purposes of examination and testing.

Soil cutting Deep excavation usually for basements, footings, or to substantially reduce existing levels.

Soil depth Extent of a *soil profile* to the *water table* or to bedrock. In undisturbed natural soil the depth is influenced by the processes of formation, with rooting volume determined by favourable aeration as opposed to the overall depth of the profile. In the urban landscape the introduction of soils and changes to existing profiles have a significant impact on the growth and *stability* of a tree, generally requiring 450–600 mm soil depth to provide sufficient moisture storage in humid areas while drier areas require a greater depth. A well drained to moderately well drained soil of 800–900 mm depth provides sufficient rooting volume, moisture storage and better mechanical support whereas shallow soils in wet conditions render trees vulnerable to *windthrow* due to lack of mechanical support (Craul 1992, p. 32).

Soil erosion Wearing away of soil physically and by chemicals in solution where the material is transported away by wind, ice or rain.

Soil fertility The amount and type of elemental nutrients and their concentrations in the soil available for plant growth considered as *macronutrients* and *micronutrients* in order of the amount of each group required to sustain plants.

Soil heave See *Heave.*

Soil horizons Distinct bands or layers of soil evident in some *soil profiles.*

Soil microflora See *Fungus.*

Soil organic matter Remains of plants and animals at various stages of decomposition, the cells and tissues of soil organisms, and substances such as *humus* made by these organisms (Handreck & Black 2002, p. 22).

Soil profile Exposed section of a soil body, usually vertically by excavation, erosion or by a *soil core sample.*

Soil stripping Reduction of soil levels by excavation.

Soil structure The presence of small aggregates formed into crumb-like clusters or *peds*, their arrangement and the pore spaces. A soil comprised mainly of peds when moist is considered *pedal* and *apedal* if peds are not apparent.

Soil texture The distribution of different sized particles in *soil*, i.e. clay, loam and sand (Handreck & Black 2002, p. 8).

Soil water See *Underground water.*

Solar access The availability of or access to unobstructed direct sunlight (United States Department of Energy 2003).

Solum The A and B horizons of a soil body (Craul 1992, p. 56).

Solution Two or more substances mixed together and uniformly dispersed in liquid, e.g. mineral salts in ground water.

Sonic detectors Electronic devices that measure the transmissibility of sound waves through a tree detecting differences based on the density of *sound wood*, deteriorated wood and cavities (Nicolotti & Miglietta 1998, pp. 297–302). Examples of such proprietary devices are the Metriguard Stress-wave Timer®, IML Impulse Hammer® and Arborsonic Decay Detector®.

Sounding Tapping of roots, trunk or branches with a mallet or hammer to sample the *acoustic resonance* to compare soundwood with wood that is decayed or *hollow.*

Sound wood Wood unaffected by *decay* or *structural deterioration.*

Sparse crown Reduced leaf density in the crown, often a precursor to *dieback* and may imply *stress* or *decline*. This may also occur as a response to drought, root damage, insect damage, herbicide or toxicity.

Species specific mulch *Organic mulch* that is derived entirely from a particular plant species. Often used as mulch to treat the same species of tree, replace nutrients as the mulch decomposes and to entice beneficial natural predators.

Specimen planting See *Specimen tree*.

Specimen tree A tree planted, retained, or occurring usually as an isolated feature and not part of a *stand* promoting its characteristics as an individual, e.g. dimensions, *crown form*, crown shape, intrinsic uniqueness *in situ*, foliage texture, flowers etc.

Spinney A small *stand* of trees.

Spiral grain See *Helical grain*.

Spiral of decline Gradual tree deterioration usually associated with repeated bouts of *predation* and *stress*. The tree exhibits increasingly *low vigour* after each *episode* until reduced *resistance* and exhaustion of reserved energy results in its demise and death.

Spiritual ownership A perception of connectedness and possession of a place or thing without necessarily being the legal owner. Often used in the context of attempting to assert influence or control over the *development* or use of a place contrary to the legal rights associated with its possession.

Sporophore The fruiting body of a fungus. May be soft or woody, *ephemeral* or persistent and may appear on the trunk, branches or roots or appear at the ground surface when the infested root is present below.

Spread See *Crown spread*.

Spring wood See *Early season's wood*.

Sprout See *Shoot*.

Sprout mass A cluster of *epicormic shoots*.

Spur A short stem usually with shortened internodes producing generally a higher number of flower buds. Such trees produce more flowers and fruit than trees without such shoots (Harris *et al.* 2004).

Stability Resistance to change especially from *loading* forces or physical modifications to a tree's growing environment. See also *Viability* and *Structural integrity*.

Stag See *Stag-headed*. See also *Snag*.

Staged cutting See *Phased target pruning* and *Pollarding*.

Stag-headed Protruding dead branches above the live foliage of the crown as a result of *dieback*. See also *Secondary crown* and *Dieback wound*.

Stag heading See *Lopping*.

Stand A group of trees often of the one kind.

Standard A technical standard is an established norm or requirement, usually as a formal document establishing uniform engineering or technical criteria, methods, processes and practices (Wikipedia 2008), e.g. Australian Standard AS 4373 Pruning of amenity trees.

Star shakes A pattern of longitudinal splitting in wood, radiating along *vascular rays* like spokes of a wheel when viewed in cross-section (Lonsdale 1999, p. 322).

State environmental planning policy (SEPP) In the State of New South Wales, Australia a *statutory plan* prepared and administered by the State Government that addresses matters of State significance.

Statement of agreed facts A document produced by *expert witnesses* after *expert witness conferencing* detailing the matters agreed upon and those remaining in dispute in a legal matter to be presented to a court at a *hearing* or trial.

Statement of contentions A written list of matters or concerns in dispute between opposing sides in a legal case requiring resolution by a court of law.

Statement of environmental effects A document supporting a *development application* by describing the proposal in detail, identifying likely impacts and the means to be used to overcome those impacts.

Statement of evidence Written document of stated facts, opinion, *policy* or presentation submitted to a court of law by an *expert witness* from a particular profession to assist the court with the making of its determination in the subject case.

Statement of issues See *Statement of contentions.*

Static leaning A leaning tree whose lean appears to have stabilised over time.

Static load Loading force that is stationary and unchanged or at rest for long periods (James 2003, p. 166). See also *Dynamic load* and *Rest point.*

Static loading See *Static load.*

Statutory law Law/s developed by parliament or legislature taking precedence over *Common law.*

Statutory plan A plan having legal status established under legislation.

Stele Comprised of *vascular tissue* such as *pith* and *vascular rays* forming a cylindrical core at the centre of stems and roots.

Stem A botanical term to describe an elongated above ground part of a plant supporting buds and leaves and containing chlorophyll, e.g. trunk and branches.

Stem bark ridge See *Branch bark ridge.*

Stick Thin branch, often dead and without foliage and lateral branches, either attached or detached from a tree.

Stilt roots Woody *prop root* or *columnar root* formed by some trees for support derived from *aerial roots,* e.g. *Ficus columnaris.*

Stop work order A *court order* for works on a site to cease.

Strain In mechanics, the distortion in an object caused by a *stress* (Lonsdale 1999, p. 322).

Strain meter An electronic device attached to a *stem* of a tree to measure stretching of fibres caused by a *dynamic load* from wind to test for *wind-firmness* (James 2005, p. 3).

Strangler roots Adventitious roots that grow from a seed attached to a host, usually in the crown, growing down to the ground becoming thickened and by *anastomosis* spreading around the host for support eventually encircling and constricting its growth causing death, e.g. *Ficus virens* and *F. macrophylla*.

Street tree 1. A tree located within the area of the *road* or *road reserve*. A tree either planted, self-sown or from *remnant vegetation,* e.g. a) A tree located in the road shoulder, b) A tree located in a median strip or a roundabout, or planting structure within the roadway. 2. A tree located within the *road reserve*. A tree either planted, self-sown or from remnant vegetation, e.g. a) Planted formally by a *consent authority* such as a local government or a roadway authority as a single or group specimen, and usually located within 1 m from the face of the kerb where the kerb and gutter is formed, b) A specimen planted as a garden activity by an adjacent property occupant, c) A self-sown specimen.

Street tree inventory See *Tree inventory.*

Strength The ability of a tree or its parts to resist *loading stress,* e.g. *compression, tension* and *torsion.*

Stress 1. A factor in a plant's environment that can have adverse impacts on its life processes, e.g. altered soil conditions (compaction, poor nutrition, and reduced oxygen or moisture levels), root damage, toxicity, drought or waterlogging. The impact of stress may be reversible given good arboricultural practices but may lead to plant *decline.* 2. In mechanics, force acting on an object, measured per unit area of the object (Lonsdale 1999, p. 322).

Stress notch See *Notch stress.*

Stress-wave timer See *Sonic detectors.*

Striker roots See *Sinker roots.*

Stringy white rot See *White rot.*

Strip footing Usually a continuous steel re-enforced concrete *footing* located along a trench beneath a load-bearing wall.

Structural branches First order or other *orders of branches* elongated to form a permanent framework of branches supporting the *crown*, persisting beyond the tree's maturity.

Structural defect A weak point in or on a tree causing its *structural deterioration* diminishing its *stability* in full or part.

Structural deterioration The diminished ability of a load-bearing part of a tree, e.g. trunk, branch, or root, to sustain its resistance to load-bearing forces under normal conditions due to wounding, breakages or decay deteriorating its *stability* in full or part.

Structural engineer An expert in the design of built structures, their performance and evaluation.

Structural failure Result of *structural deterioration* where a tree or tree part succumbs permanently to superior *loading* forces resulting in collapse in full or part of trunk, branch, or root.

Structural integrity The ability of a load-bearing part of a tree, e.g. trunk, branch or root under normal conditions to sustain its resistance to *loading* forces.

Structural root zone (SRZ) The minimum radial distance around the base of a tree and its *root plate* required for its stability in the ground against *windthrow*, and applied only to trees with a circular root plate (Mattheck & Breloer 1994, pp. 77–87).

Structural roots Roots supporting the infrastructure of the *root plate* providing strength and *stability* to the tree. Such roots may taper rapidly at short distances from the *root crown* or become large and woody as with gymnosperms and dicotyledonous angiosperms and are usually first and second order roots, or form an *adventitious root mass* in monocotyledonous angiosperms (palms). Such roots may be crossed and grafted and are usually contained within the area of *crown projection* or extend just beyond the *dripline*.

Structural soil A plant media usually of rock of a given diameter to provide large interlocking spaces filled with soil and added nutrients into which a tree can be planted allowing for root growth beneath a surrounding load-bearing area such as a footpath, driveway or road.

Structural woody roots See *Structural roots.*

Structural wound Any wound occurring on a tree as a result of a *structural failure,* e.g. branch splitting or *hazard beam*, diminishing its *stability* in full or part.

Structurally deteriorated See *Structural deterioration.*

Structure See *Soil structure.*

Stub cut See *Lopping.*

Stump Remaining section of *trunk* attached to the *root plate* after the *crown* of a tree has failed and *collapsed* or has been *removed.*

Stump grind The application of a grinding device to remove a stump and usually the *root crown* and part of the *root plate* to various depths below ground and varying distances from the trunk.

Stump grinder A device to remove a tree's stump by a grinding process.

Sub-base Introduced material compacted as a *foundation* for building, e.g. gravel.

Suberin In bark, a fatty substance present in the cell walls that functions to prevent or resist *decay* entry and water loss.

Sub-grade *Extant* lower soil levels compacted as a *foundation* for building works.

Suboptimal calliper Stem thinning forming *hour glass* as a result of *palm over-pruning* (Harris *et al.* 2004, p. 381).

Subordinate Branch or tree suppressed by another. See also *Crown form suppressed.*

Subordination Pruning usually in *young* trees of *deliquescent* habit to develop dominant *structural branches* or to slow growth by reducing stem length or the selective removal of *laterals* of proportionate diameter, usually one-half

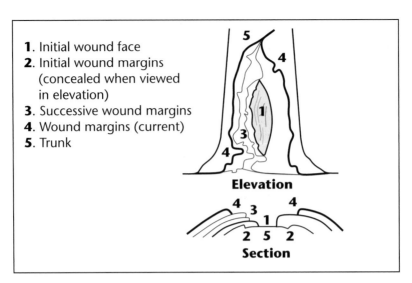

Figure 26 Successional wound.

(½) to rarely three-quarters (¾) that of the stem being removed (Gilman 1997, pp. 13, 28).

Subpoena A *court order* for an individual to attend a court at a prescribed time and date to give *evidence* or testimony or to produce a deposition. Failure to comply with a subpoena may lead to prosecution.

Subsidence crack A crack occurring on the underside of an upwardly curved stem, which is not *hollow*, and is bent in the opposite direction to the curve as a result of excessive *loading forces* causing initially the *vascular rays* to split as they are pulled at right angles to their longitudinal orientation (Mattheck & Breloer 1994, pp. 166–169).

Substantial A tree with large dimensions or proportions in relation to its place in the landscape.

Successional wound Preceding layers of failed wound margin/s forming a step-like sequence away from the *wound face*, where present, to the current wound margin/s indicating repeated cycles of formation and failure of *CODIT Wall 4*. (See Figure 26.)

Sucker *Epicormic shoot* growing from a *latent bud* in older wood. Such shoots are vigorous and usually upright and arise below the graft union on the *understock* or at or below ground from the trunk or roots (Harris *et al.* 2004, p. 18). See also *Watershoot.*

Sucker shoot See *Sucker.*

Sudden branch drop The *failure* and *collapse* of live, usually horizontal branches, seemingly without any noticeable cause in calm hot, dry weather conditions generally after rain. Theorised to be caused by altered moisture content in the branch disturbing the longitudinal *prestressing* of the wood that normally helps support the load as formed by *reaction wood* in branches tending to horizontal (Lonsdale 1999, p. 30), or *incipient failure* from the lengthening of existing internal cracks as the wood cools (Shigo 1986, p. 248), or influenced by *branch creep* under its own weight and by wind (Mattheck & Breloer 1994, p. 126), or fractures to *vascular rays* if pulled at right angles to their longitudinal orientation forming from *subsidence cracks* (Mattheck & Breloer 1994, p. 169), or a combination of these factors. Such branch breakages usually occur at some distance from the branch collar leaving a stub. See also *Branch tear out.*

Sudden limb drop See *Sudden branch drop.*

Sudden limb shear See *Sudden branch drop.*

Sugars Water soluble food storage carbohydrates (Lonsdale 1999, p. 323).

Summer branch drop See *Sudden branch drop.*

Summer wood See *Late season's wood.*

Sunken spot An area of depression on a stem or trunk below a branch where a *branch collar* appears absent but has formed further down the stem or branch (Shigo 1991, p. 86), or an area of depression below a branch that has died or been removed and the tissue below the branch subsequently wanes (Shigo 1989a, p. 439).

Sun scald Wounding usually on the upperside of branches after sudden exposure to sunlight especially in summer, e.g. after excessive pruning of the upper crown, or following storm damage stripping foliage or branches, e.g. *Ficus* spp.

Superior Of *First order structural branches* (FOSB) the branch or branches having the largest diameter and generally supporting singly the greater percentage of *crown cover* usually asserting *apical dominance* suppressing other smaller *First order structural branches* and *lateral branches*. This may vary on a tree that has had its crown structure modified. See also *Inferior*.

Suppressed bud See *Latent bud*.

Surgery See *Tree surgery*.

Survey marker wound See *Blaze*.

Survey plan A plan depicting contours, boundaries and existing features.

Suspended slab A concrete slab supported by piers or bridge footings usually with a void beneath.

Sustainable retention index value (SRIV) A visual tree assessment method to determine a qualitative and numerical rating for the viability of urban trees for development sites and management purposes, based on general tree and landscape assessment criteria using classes of *age, condition* and *vigour*. SRIV is for the professional manager of urban trees to consider the tree *in situ* with an assumed knowledge of the *taxon* and its growing environment. It is based on the physical attributes of the tree and its response to its environment considering its position in a matrix for age class, vigour class, condition class and its sustainable retention with regard to the safety of people or damage to property. This also factors the ability to retain the tree with remedial work or beneficial modifications to its growing environment or removal and replacement. SRIV is supplementary to the decision made by a tree management professional as to whether a tree is retained or removed (Institute of Australian Consulting Arboriculturists 2005).

Sweep See *Butt sweep*.

Symbiotic An association between different species usually but not always mutually beneficial.

Symmetry Balance within a *crown,* or *root plate,* above or below the *axis* of the trunk of branch and foliage, and root distribution respectively and can be categorised as *asymmetrical* and *symmetrical* (Figure 27).

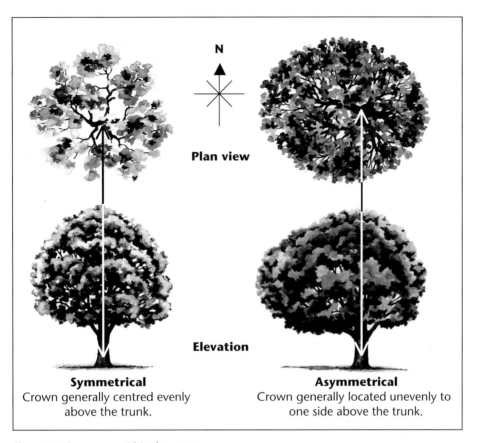

Figure 27 Symmetry within the crown.

Symmetrical Balance within a crown, where there is an even distribution of branches and the *foliage crown* around the vertical *axis* of the trunk. This usually applies to trees of *crown form dominant* or *crown form forest*. An example of an expression of this may be crown symmetrical. See also *symmetry* and *asymmetrical*.

Symmetrical cavity A *cavity* formed with a generally even radial development from the *axis* of a stem or root. See also *Asymmetrical cavity*.

Symmetrical decay An area of *decay* formed with a generally even radial development from the *axis* of a stem or root. See also *Asymmetrical decay*.

Sympathetic removal Removal of tree/s by a method that would seem the least harsh to the tree and cause minimal disturbance to the physical environment and its amenity, especially in a location where the works are prominent and exposed to the public where perceptions of such works may be readily misunderstood, e.g. the lesser impact of the removal of a stand of trees disassembling them by chainsaw and grinding out individual stumps, as opposed to higher impact and harsher approach of displacing the stand with excavation equipment, clearing all other vegetation and topsoil and piling the trees and other material into windrows.

Symplast Living plant tissue, connecting adjacent cells by fine strands (plasmodesmata) which pass through pits in the cell wall. This is centred in the *vascular cambium* and is the network of live tissue in wood and bark (Lonsdale 1999, p. 323; Shigo 1986, p. 110).

Systematic removal Trees removed in a planned manner, e.g. during a *planned maintenance* program of tree works. See also *Alternation of generation*.

Tangential A plane which intersects a radial plane (e.g. that of a xylem ray) at right angles (Lonsdale 1999, p. 323).

Tangential core An *increment core* extracted across a *stem* in a line parallel to the centre of the stem (Weber & Mattheck 2003, p. 44).

Tangential pressure On straight trees a *tension* force at the surface of a stem exerted radially, evident as broader spindle shaped *vascular rays* (Mattheck & Breloer 1994, pp. 168–169).

Taper In roots and branches the decrease in diameter along a given length, usually reducing gradually in the *distal* direction (Lonsdale 1999, p. 323).

Taproot Where the *radicle* persists as the initial descending root in some trees and remains of structural importance. The taproot may be replaced or reduced in importance by the development of *lateral roots, buttress roots, heart roots* and *sinker roots*. Some taproots descend to the lowest level of the *root plate* where they terminate and may give rise to lateral roots. The extent of the depth of such vertical roots may be restricted when oxygen levels are low or they are obstructed or reach the *water table* (Thomas 2000, pp. 73–74).

Taprooters Trees with a *root plate* morphology where a *taproot* develops deep into well-drained soil for *anchorage* and *lateral roots* form for *stability*. Such trees may be subject to *windthrow* where the soil cracks and lateral roots break allowing a *root ball* to rotate in the ground causing the taproot to snap at the edge of the *root ball* (Mattheck & Breloer 1994, pp. 75–77).

Taprooting Trees that develop a *taproot* to resist loading forces but this is dependent on the compressive resistance of the soil. Such trees usually form *sinker roots* for *stability* and additional support against *windthrow* (Craul 1999, p.166).

Target People or property likely to be harmed or damaged, respectively, by being struck by a failed or collapsed tree in full or part.

Target canker See *Canker.*

Target cut See *Final cut.*

Target pruning Pruning of selected branches that pose a hazard or *nuisance* or may grow to create a hazard, e.g. *crossed branches*, branches growing over a dwelling or aerial utility cables.

Taxa Plural of taxon, see *Taxon.*

Taxon A taxonomic group of any rank, e.g. family, genus, species, variety or cultivar.

Tear out See *Branch tear out.*

Tear out wound A wound of usually concave shape created by a *branch tear out.*

Temporary branches Branches sustained by a tree for a short period of time. Such may be found on trees in a stand, or forest environment, overshadowed or restricted for space and light by competition or by its own crown or other trees or structures and gradually shut down, compartmentalised and abscised by natural pruning. Where competition causes the tree to grow tall up to the light the non-structural branches along the trunk no longer supporting the rapid extension of the crown become transitional and isolated by the elongation of the typically straight trunk when asserting *apical dominance* or elevation to the upper canopy, and may become overshadowed and inefficient. Such branches may be divergent, ranging from horizontal to ascending and are usually evident in the lower crown and along the trunk. See also *Non-structural branches* and *Permanent branches*.

Temporary tree protection zone fence See Tree protection zone fence.

Temporary roots A root that may live for a short period of time before declining or dying back, being replaced by new roots branching nearby, e.g. a *radicle* may persist or be replaced by adventitious or *lateral roots* as with palms.

Temporary wilting point The short-term loss of *turgor pressure* in plant cells when *soil water* is unable to be taken up by roots. A plant may recover with access to available *soil water* or from reduced transpiration, e.g. by being shaded. With continued *episode*s of this occurrence the plant will *decline*, evident by the abscission of leaves and drying of other plant parts possibly resulting in *premature aging* or *irreversible decline*. See also *Wilting, Wilting point* and *Permanent wilting point*.

Tensile Loading force from *tension* forces.

Tensile buttressing Narrow *buttressing* of *first order roots* on tropical rainforest trees with the *roots* surrounding the trunk usually in all directions providing support in *tension* but with little compressive value (Craul 1999, p. 166).

Tension The action of being stretched. Evident on the upperside of a stem or root usually exhibited as a smooth surface along the path of the bending force. This may be evident by loose or popping bark. See also *Tension wood*.

Tension wood *Reaction wood* formed in dicotyledonous *Angiosperms* as additional wood growth on the upper side of a stem opposing a lean, reacting to the loading stimulus to pull the stem upwards.

Terminal bud The bud growing at the end of a branch or stem. See also *Apical meristem*.

Termite leads Tunnels of *mud* on the stem and between bark created by termites that may be active or inactive.

Texture See *Soil texture*.

Thermotropism A plant growth response to temperature stimulus.

The site Land or premises subject of arboricultural works, especially when referred to repeatedly in an arboricultural report, e.g. a property subject to building *development* affecting trees being 3 Smith Street (*the site*).

Thicket A dense growth of trees and shrubs.

Thigmomorphogenesis The *adaptive wood* response of exposed trees or *edge trees* to wind loading by being shorter, having twisted shorter trunks, and thicker stronger roots to prevent failure (Ennos 2001, p. 45).

Thigmotropism A directional growth movement of a plant or part by bending or turning after the stimulus of contact with a solid object or surface.

Thinning See *Crown thinning.*

Thinning cuts See *Crown thinning.*

Tipping See *Reduction pruning.*

Tissue A group of cells of the same type having a common function.

Tomograph A disc-like sectional image taken through an object, used in trees to detect the extent of *decay* or a *cavity* in a root or stem.

Tomography Any process producing a *tomograph.*

Top See *Topping.*

Topiary The art and practice of forming plants into desired shapes by pruning and training branches.

Top lopping See *Topping.*

Topping Removal of the upper part of a tree, reducing its height by *lopping.* This practice usually damages trees, reducing strength, condition and vigour promoting *premature decline* and exposure to pests and diseases. See also *Lopping.*

Top soil stripping See *Soil stripping.*

Torsion The action of being twisted. Wood fibres arranged along a spiral force flow.

Torsional stress Loading force from *torsion* forces.

Town planner A professional who designs and plans the layout of towns and cities and administers planning policies for different levels of government.

Toxic Poisonous.

Toxin Any natural or artificial substance present in or introduced into the environment in quantities or concentrations poisonous to a tree.

Trace elements See *Micronutrients*.

Tracheids In *gymnosperms*, vertically aligned tubes comprised of single dead cells for transporting liquids in *xylem*.

Tracing See *Scribing*.

Translocation The movement of nutrients and *sugars* in solution from one cell to another.

Transmission easement See *Right of way*.

Transpiration The release of moisture as vapour to the atmosphere from the surface of a plant, especially via the stomata of leaves.

Transplant See *Transplanting*.

Transplantation See *Transplanting*.

Transplanters Individuals or business specialising in *transplanting* trees.

Transplanting Extraction of a tree from its place within the ground or fixed containerised position to relocate it for re-establishment within the ground or another containerised area.

Transport roots See *Outer roots*.

Transverse stress A loading force at a right angle to a structure, e.g. such as causes a *hazard beam*.

Trap See *Root barrier* and *Trapping*.

Trapping Root barriers comprised of porous sheets with fine gaps ≤1 mm to allow for root tip elongation, but strong enough to cause root pruning by strangulation and girdling as radial growth expands. Such barriers may be constructed of welded metal, fibre sheets or woven or non-woven fabrics (Coder 1998, p. 63; Roberts *et al.* 2006, p. 355).

Tree A woody *perennial* plant long-lived, greater than, or potentially greater than, 5 m high with one trunk and one or relatively few stems. See also *Shrub*.

TreeAZ A method of tree assessment developed by Jeremy Barrell in 2002 to categorise the relative importance of trees on development sites and as a replacement that evolved from the *Safe Useful Life Expectancy* (SULE) and British Standard BS 5837:1991 methods of assessing trees. TreeAZ was designed around the principles of tree management to reduce risk and to sustain amenity (Barrell 2002).

Tree audit See *Tree inventory*.

Tree injection Application of systemic pesticides or nutrients in liquid form to the vascular cambium by injection.

Tree inventory Detailed list of information about trees in a specific area, such as tree location, species, physical dimensions, health/*vigour, condition, age class* etc. Information is obtained by survey and location may be established by a global positioning system. A tree inventory may assist with an asset management program.

Tree management Planned protection, conservation, maintenance and enhancement of a population of trees. Usually achieved by recognising trees as a dynamic natural resource and, through professional arboricultural personnel and a multidisciplinary approach, gaining an ongoing understanding of diverse aspects of the population: age class; maintenance, removal/replacement cycles and costs; additional new planting opportunities and costs; sustainability; safety constraints; community concerns; budgetary constraints; ecological, *amenity* and utility values; suitability and appropriateness of tree maintenance, removal and replacement or retention. See also *Tree preservation, Appropriate tree management* and *Inappropriate tree management*.

Tree measuring tape See *Diameter tape*.

Tree preservation Protection, conservation and maintenance of a population of trees. See also *Tree management*, *Appropriate tree management* and *Inappropriate tree management*.

Tree preservation order (TPO) An ordinance made under planning *legislation* to protect trees generally, or specifically for their importance for amenity, heritage, landscape, environmental and nature conservation.

Tree protection zone (TPZ) A combination of the *root protection zone (RPZ)* and *crown protection zone (CPZ)* as an area around a tree set aside for the protection of a tree and a sufficient proportion of its growing environment above and below ground established prior to demolition or construction and maintained until the completion of works to allow for its viable retention including *stability*. Generally this will be primarily the RPZ and to a lesser extent the CPZ. This may be delineated as a specifically fenced-off area around the tree at a distance from the trunk determined as multiples of the trunk diameter (DBH) and dependent on age class and vigour class. Special protection or construction works may provide a tree protection zone without a fence having been erected, e.g. barriers formed by site sheds located on piers. Such a protection area may form an exclusion zone for all works including the temporary or permanent location of utility services. Note: Any *encroachment* into the area would require additional tree protection specifications or works in consultation with the *Site arborist*.

Tree protection zone fence The fence constructed to define a tree protection zone or that part of the TPZ specifically fenced off. A typical TPZ fence would be 1.8 m high steel chain link mesh with galvanised steel pipes, with the perimeter further delineated by the attachment of shade cloth material to the outside surface area of the fence to reduce the movement of dust and other airborne residue from building activities that may be phytotoxic. A 'Tree Protection Zone' sign restricting access would typically be affixed to the fence. Such a fence should be installed prior to the commencement of any works on site, except weed removal and tree maintenance, e.g. pruning, irrigation and mulching, and should be maintained for the duration of the project. A lockable opening for access is desirable to secure the enclosure. A

TPZ fence or fence section may be temporarily established initially where demolition of existing structures is required to provide an area of sufficient space for the full extent of the Tree Protection Zone to be installed at a later stage.

TreeRadar™ See *Tree Radar^{TM} Unit*.

Tree Radar™ Unit (TRU™) Proprietary system of *ground penetrating radar* utilising the same non-destructive technology to internally examine and analyse the extent of *decay* or hollows within a *stem* (TreeRadar Unit Inc. 2007).

Tree removal See Removal/remove.

Tree spade A mechanical device, usually truck mounted, used for *transplanting* trees, especially palms.

Tree stump See *Stump*.

Tree surgeon See *Arborist*.

Tree surgery *Pruning* undertaken to a recognised Standard, or remedial works performed to maintain or prolong the life of a tree.

Tree survey See *Tree inventory*.

Tree transplanters See *Transplanters*.

Tree valuation A process to determine the monetary or intrinsic worth of a tree.

Tree worker See *Arborist*.

Trenchless technology Methods or systems that allow for the installation, replacement, renovation and repair of pipes, ducts, cables and other underground apparatus with minimum excavation from the ground surface. These include horizontal directional drilling (HDD), the most common system used for new pipe installation in domestic situations is pipe bursting (pipe cracking) and the most common system used for pipe replacement in domestic situations are: soil displacement hammer, pipe ramming, thrust boring, micro tunnelling and pipe lining (Australasian Society for Trenchless Technology 2003).

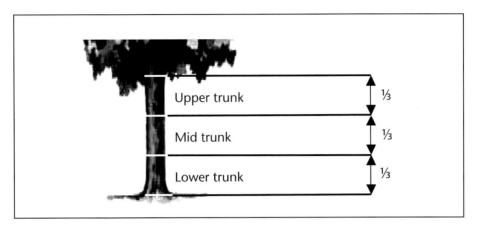

Figure 28 Trunk sections.

Tropic See *Tropism*.

Tropism A growth response to a stimulus. Directional growth movement of a plant or plant part that occurs in response to an external stimulus from a specific direction, being either positive or negative depending on whether movement is towards or away from the source of stimulation.

Trunk A single stem extending from the *root crown* to support or elevate the *crown*, terminating where it divides into separate *stems* forming *first order branches*. A trunk may be evident at or near ground or be absent in *acaulescent* trees of *deliquescent* habit, or may be continuous in trees of *excurrent* habit. The trunk of any *caulescent* tree can be divided vertically into three sections and can be categorised as *lower trunk*, *mid trunk* and *upper trunk*. For a *leaning* tree these may be divided evenly into sections of one-third along the trunk (Figure 28).

Trunk circumference Distance around the outside of the trunk. Also calculated as π(Pi) (3.142) × diameter (*DBH*) using the formula for circumference (Farm Forest Line 2002).

Trunk collar See *Branch collar*.

Trunk diameter at breast height See *Diameter at breast height (DBH)*.

Trunk diameter circular Measurement of trunk width where the distance from the centre of the trunk to the bark is equidistant, recorded using calipers or a measuring tape or *diameter tape.*

Trunk diameter non-circular Measurement of trunk width, where the distance from the centre of the trunk to the bark is not equidistant, recorded using calipers or a tape, recording the longest and narrowest widths to calculate an average from the two measurements (Farm Forest Line 2002).

Trunkless A tree without a trunk, *acaulescent.* Usually branches immediately above the *root crown*, e.g. mallee Eucalypts, *Lagerstroemia indica* – Crepe Myrtle. In palms, a trunkless species usually has an underground trunk (Jones 1996, p. 272).

Turgid Swollen and firm with water due to pressure.

Turgor See *Turgor pressure.*

Turgor pressure The pressure developed in a cell as it becomes filled with water.

Twig Lowest order of branching in most trees usually containing the recent or *current season's growth* and mainly concentrated in the *outer extremity* of the crown.

Tyloses Plural of *tylosis.* See *Tylosis.*

Tylosis A balloon-like extension of a xylem parenchyma cell into an adjoining vessel via the pits in the adjoining cell wall. Tyloses assist in sealing *dysfunctional xylem* (Lonsdale 1999, p. 324).

Typical Having characteristics representative of others in a taxonomic group, e.g. of a species. In a tree this may also be growth that is representative for *crown form, habit* and type or behaviour expected to occur naturally. See also *Atypical* and *Misshapen.*

Ultrasonic detectors 1. Electronic devices that measure the transmissibility of *ultrasound* waves through a tree detecting differences based on the density of *sound wood*, deteriorated wood and cavities (Nicolotti & Miglietta 1998, pp. 297–302). Examples of such proprietary devices are the Silvatest® and Arborsonic®. 2. Ultrasonic *tomography*. Electronic devices that measure the transmissibility of ultrasound waves through a given section of a tree by detecting pulses through multiple sensors placed around a *stem* to indicate areas with the same density (Nicolotti & Miglietta 1998, pp. 297–302). An example of such a proprietary device is the Picus® *Tomograph*.

Ultrasonic tomography See *Ultrasonic detectors*.

Ultrasound The frequency of sound above the limit of human hearing, i.e. from approx. 20 kHz upwards (Lonsdale 1999, p. 324).

Underground water Water held below ground in soil and permeable rocks.

Understock Root system and stem section of a plant known to be vigorous and disease resistant to support the *graft* of a *scion*. Used to propagate a cutting of a plant variety or cultivar usually of the same genus but chosen for its specific horticultural characteristics, e.g. *crown form*, fruit, flower or foliage colour.

Undesirable See *Undesirable species.*

Undesirable species Plants that have characteristics which may be harmful as a result of toxin production and potential poisoning, potential for weed infestation, brittle and dangerous wood, excessive spread of roots or bushland invasion.

Undesirable trees See *Undesirable species.*

Unplanned maintenance Tree maintenance undertaken as part of unscheduled works usually in response to requests or episodes of failure and collapse of a tree or tree part. See also *Planned maintenance, Tree management* and *Tree preservation.*

Upper crown The *distal* or highest section of a crown when divided vertically into one-third (⅓) increments. See also *Crown, Mid crown* and *Lower crown.*

Upper trunk Highest, or *distal* section of a trunk when divided into one-third (⅓) increments along its *axis.* See also *Trunk, Lower trunk* and *Mid trunk.*

Uprooted A tree where the *root plate* is dislodged from the ground. See also *Windthrow.*

Urban community forestry See *Urban forestry.*

Urban forestry Management of the entire population of trees and woody shrubs in an urban environment recognising them as critical elements of urban *infrastructure* providing physiological, sociological, economic and aesthetic benefit. See also *Tree management, Tree preservation* and *Arboriculture.*

Utility services The services located either above or below ground that may impact upon the growth and management of trees and may be delivered by pipes, cables or radio waves. Many of these services are located within the *road reserve.*

Vacuole A fluid-filled sac within a cell.

Vascular bundle A strand of conducting tissue containing *xylem* and *phloem*.

Vascular cambium A lateral meristem of conductive cells which gives rise to *secondary xylem* and *secondary phloem*.

Vascular ray A narrow sheet of *parenchyma* cells running radially across the secondary vascular tissues of a stem or root used to translocate materials in solution and nutrients laterally, providing wood strength and particularly radial support.

Vascular tissue A group of food or water-conducting cells.

Vegetation management plan A structured program for the protection, maintenance, restoration and replacement of trees and native vegetation.

Vehicular crossover Section of the *road reserve* utilised to construct a driveway to an adjoining property and usually connected to the roadway by a *layback*.

Vein A strand of *xylem* and *phloem* in a leaf blade.

Verge See *Road reserve*.

Vertical mulching A process to reduce *soil compaction* and improve movement of air, water, nutrients and roots in soil by drilling, or use of hydraulic or pneumatic probes to create holes of 50 mm diameter to 450 mm deep at 1 m spaces replacing the soil removed with coarse material to prevent collapse (Kalisz, Stringer & Wells 1994, pp. 141–145). See also *Radial trenching*.

Vessels In angiosperms, vertically aligned tubes comprised of numerous dead cells for transporting liquids in *xylem*.

Veteran tree A senescent *significant tree* especially one seeming to have survived beyond the *typical* life expectancy for the *taxon*. The survival of such trees may be *prolonged senescence* attributed to human intervention.

Viability 1. Estimate of the remaining life span of a tree with consideration of its *age*, *condition*, and impact on safety. 2. The remaining time left that seeds are likely to successfully germinate. See also *Stability* and *Life expectancy*.

Vigour Ability of a tree to sustain its life processes. This is independent of the *condition* of a tree but may impact upon it. Vigour can appear to alter rapidly with change of seasons (seasonality), e.g. *dormant*, deciduous or semi-deciduous trees. Vigour can be categorised as *normal vigour*, *high vigour*, *low vigour* and *dormant tree vigour*.

Visual tree assessment (VTA) A visual inspection of a tree from the ground based on the principle that, when a tree exhibits apparently superfluous material in its shape, this represents repair structures to rectify defects or to reinforce weak areas in accordance with the *axiom of uniform stress* (Mattheck & Breloer 1994, pp. 12–13, 145). Such assessments should only be undertaken by suitably competent practitioners.

Vitality See *Vigour*.

Vitamat® See *Electrical conductivity meter.*

Volunteer top See *Asserted dominance.*

Water demand Apparently different water requirements among *taxa* influenced by *age* and *height* and can be categorised as high, medium or low (O'Callaghan & Lawson 1995, p. 104) or 'the amount of water required by a tree in order to keep its metabolism functioning at optimum levels to meet its physiological requirements' (Lawson & O'Callaghan 1995, pp. 90–97).

Water knife A hydraulic device that uses a fine stream of water pumped under sufficient pressure to displace soil or cut roots. At lower pressure soil may be displaced only, allowing woody roots to be exposed for examination or *root mapping*. See also *Air knife*.

Water logging Inability of a soil body to drain freely and support root growth. This may result in *leaning* or *windthrow failure*. Exceptions are trees adapted with modified roots such as mangroves.

Watershoot *Epicormic shoot* growing from a *latent bud* in older wood. Such shoots are vigorous and usually upright and arise above the ground or graft union on the *scion* or from the trunk or older branches (Harris *et al.* 2004, p. 16). See also *Sucker*.

Watersprout See *Watershoot*.

Water table See *Ground water level*.

Wattle Day In Australia, a day initiated in 1838 to promote and celebrate wattle (*Acacia* spp.) as an identifying Australian floristic symbol with the national day proclaimed in 1992 as 1 September, the first day of spring in all States and Territories. Green and Gold are the National Colours of Australia (proclaimed 1984) and *Acacia pycnantha* – Golden Wattle, is the National Floral Emblem of Australia (proclaimed 1988) (Wattle Day Association 2005).

Weak junctions Points of possible branch failure in the crown, usually caused by the trunk or branch bark being squeezed or concealed within the junction so that the necessary interlocking by intertwining wood fibres does not occur and the junction is forced open by branch movement and exacerbated by the annual increments of growth in the opposing stems and accumulated bark. This may be a genetic problem with some taxa. See also *Compression fork*.

Weed A plant growing out of place or where it is not wanted and often characterised by the high production of viable seeds and their ability to quickly colonise disturbed ground. Note: Most countries and their states will have their own weeds *legislation* and here definitions may vary.

Weed species Any plant species *exotic* or *native* which is known to spread by the production of viable *progeny* often in large numbers, outcompeting and disrupting existing vegetation, e.g. in gardens, parks or bushland. The species concerned may be introduced from outside its area of natural distribution to an area where there are few or no natural predators, or it may have an ability to spread due to changes in land use creating a favourable habitat.

Welded fork A *compression fork* that has grafted with its *included bark* becoming *enclosed bark* (Mattheck & Breloer 1994, p. 60) and may or may not form a *rib*.

Welding The grafting of convergent branches in and above a crotch containing *included bark*, where it becomes enclosed (Mattheck & Breloer 1994, p. 60). See also *Enclosed bark* and *Compression fork*.

Wet wood A disease of wood caused by anaerobic bacteria, but while present prevents the development of *decay*.

Whip A young, single stemmed tree supplied for planting before the appreciable development of *lateral* branches.

White rot Decomposition by fungi of all cell wall components of wood, and here *lignin*, *cellulose* or *hemicellulose* may be attacked in different orders. Some advanced *decay* removes most of the *lignin* leaving a small residue of light-coloured cellulose often in localised pockets. White rot is further sub-divided into stringy, spongy, mottled or pocket rots and the rot is often yellow or more yellow-brown than white with the decayed wood appearing fibrous. This rot is usually caused by fungi from the subdivisions Hymenomycetes of the subdivision Basidiomycotina (Manion 1991, pp. 227–228). This decay affects the torsional strength of wood.

Widow maker A euphemism for a tree or tree part considered to be structurally unsound and therefore extremely hazardous.

Willful destruction Any deliberate process or activity inflicted to cause wounding, injury or death of a tree.

Wilting Collapse of plant cells caused by a loss of *turgor pressure* resulting in water being unavailable to the leaves. This may be caused by dry soil, root loss, salinity, flooding, root girdling, high temperatures, disruption by insects to xylem and gas leaks in the soil (Harris *et al.* 2004, p. 457). See also *Wilting point*, *Temporary wilting point* and *Permanent wilting point*.

Wilting point A marked loss of *pressure* in plant cells when *soil water* is unavailable and therefore unable to be taken up by roots. See also *Wilting*, *Temporary wilting point* and *Permanent wilting point*.

Wind exposure The degree to which a tree or other object is exposed to wind, with regard to both duration and velocity.

Windfirmness Varying characteristics of different tree taxa to resist wind damage, instability or *windthrow* (Craul 1999, p. 164).

Wind pressure The force exerted by wind on a tree.

Windrow Linear piles of *removed* trees and other plant material.

Wind snap The breaking of a tree's stem by wind.

Windthrow Tree failure and collapse when a *force* exerted by wind against the *crown* and *trunk* overcomes resistance to that force in the *root plate*, such that the *root plate* is lifted from the soil on one side as the tree tips over.

Windward The side of an object exposed to wind direction or the upper side of a lean. See also *Leeward*.

Winter bud A structure on some trees to protect growing points during dormancy over winter, comprised of a bud and enclosing *cataphylls*.

Witches broom A virus or other pathogen-initiated reaction culminating in excessive branching from a given point.

Wood A dense *tissue* comprised of living, dying and dead cells in the branches, trunk and roots providing flexibility and support. The cell walls are comprised of *cellulose, hemicelluloses* and *lignin* (Shigo 1986, p. 125).

Woodland grown See *Forest grown*.

Wood loss A decrease generally in the quantity of *wood* in a stem.

Woody roots See *Roots*.

Wound Damage inflicted upon a tree through injury to its living cells, from biotic or abiotic causes, e.g. where *vascular cambium* has been damaged by branch breakage, impact or insect attack. Some wounds *decay* and cause *structural deterioration* or *defects*. Trees of *normal vigour* are able to resist and contain infection by walling off areas within the wood by *compartmentalisation*. See *Compartmentalisation of decay in trees (CODIT)*. An *occlusion* may eventually conceal a wound but the enclosed *defect* remains internally and *decay* may continue to develop further weakening the *heartwood* and *sapwood* compromising the tree's *structural integrity*. The cause of a wound may be accidental, e.g. *branch tear out,* or deliberate, e.g. *carved tree*. (See Figure 29.)

Wound apex The *distal* end of a *wound*. The shape may be acute, irregular, jagged, obtuse, rounded or truncate (Figure 29).

Wound apex acute Apex of a wound that is tapering and the *occlusion* interface angle is less than <90°.

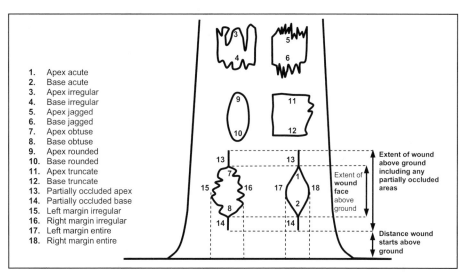

1. Apex acute
2. Base acute
3. Apex irregular
4. Base irregular
5. Apex jagged
6. Base jagged
7. Apex obtuse
8. Base obtuse
9. Apex rounded
10. Base rounded
11. Apex truncate
12. Base truncate
13. Partially occluded apex
14. Partially occluded base
15. Left margin irregular
16. Right margin irregular
17. Left margin entire
18. Right margin entire

Extent of wound above ground including any partially occluded areas

Extent of wound face above ground

Distance wound starts above ground

Figure 29 Wounds, margins, bases and apexes.

Wound apex irregular The *wound wood* growth at the apex mostly interrupted forming an edge that is not uniform or jagged. Often this may be influenced by a *successional wound* resulting in disproportionate development of *callus wood* and *wound wood*.

Wound apex jagged The *wound wood* growth or tissue damaged initially at the apex that is uneven and likely to have been caused by laceration.

Wound apex obtuse Apex of a wound that is tapering and the *occlusion* interface angle is greater than >90°.

Wound apex rounded The *wound wood* growth at the apex that is curved.

Wound apex truncate The *wound wood* growth or tissue damaged initially at the apex that is even and likely to have been caused by incision.

Wound base The *proximal* end of a *wound*. The shape may be acute, irregular, jagged, obtuse, rounded or truncate (Figure 29).

Wound base acute Base of wound that is tapering and the *occlusion* interface angle is less than <90°.

Wound base irregular The *wound wood* growth at the base mostly interrupted forming an edge that is not uniform or jagged. Often this may be influenced by a *successional wound* resulting in disproportionate development of *callus wood* and *wound wood*.

Wound base jagged The *wound wood* growth or tissue damaged initially at the base that is uneven and likely to have been caused by laceration.

Wound base obtuse Base of wound that is tapering and the *occlusion* interface angle is greater than >90°.

Wound base rounded The *wound wood* growth at the base that is curved.

Wound base truncate The *wound wood* growth or tissue damaged initially at the base that is even and likely to have been caused by incision.

Wound closure See *Occlusion*.

Wound dressing A chemical application to a *wound face* for the purposes of aiding its *occlusion*, appearance and to prevent *decay*, but rarely with any success (Harris *et al.* 2004, pp. 452–453).

Wound face Surface area of tissue exposed by injury, e.g. bark, sapwood, heartwood (Figure 30 on p. 173).

Wound face cracks horizontal Transverse cracks in a *wound face* indicative of failure from *tension* force (Mattheck & Breloer 1994, p. 183). (See Figure 31 on p. 174.)

Wound face cracks vertical Longitudinal cracks in a *wound face* indicative of failure from *compression* force (Mattheck & Breloer 1994, p. 183). (See Figure 31 on p. 174.)

Wound face entire Surface of exposed tissue is uniform without damage extending to a different layer or unaffected by borers or decay, e.g. possibly described as *wound face* entire to dead sapwood.

Wound face exposed heartwood Wound extending to reveal the *heartwood*, or has deteriorated through *decay* to reveal this layer of wood.

Wound face exposed sapwood Wound extending to reveal the sapwood, or has deteriorated through *decay* to reveal this layer of wood.

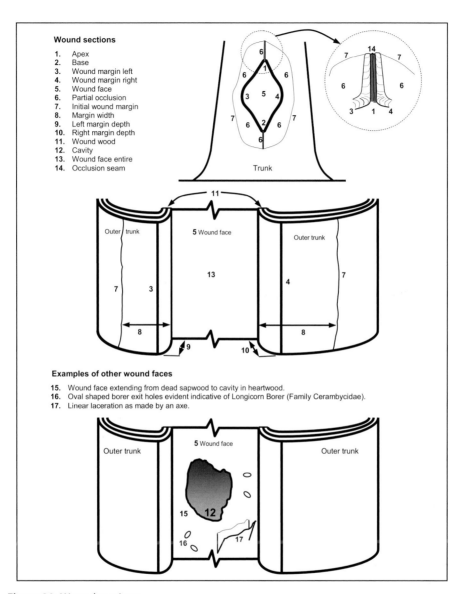

Wound sections

1. Apex
2. Base
3. Wound margin left
4. Wound margin right
5. Wound face
6. Partial occlusion
7. Initial wound margin
8. Margin width
9. Left margin depth
10. Right margin depth
11. Wound wood
12. Cavity
13. Wound face entire
14. Occlusion seam

Examples of other wound faces

15. Wound face extending from dead sapwood to cavity in heartwood.
16. Oval shaped borer exit holes evident indicative of Longicorn Borer (Family Cerambycidae).
17. Linear laceration as made by an axe.

Figure 30 Wound sections.

Wound margin The left and right sides of a *wound* as bound by the alignment of fibres along a stem or root longitudinally, being either the remaining undamaged living cells and new *callus wood* and *wound wood* on older wounds. Here the fibres are usually formed from *meristematic* cells. A wound margin

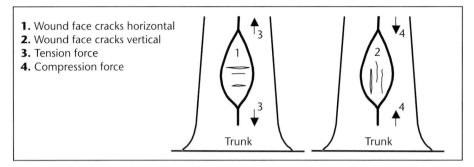

Figure 31 Wound face cracks. (Source: Mattheck & Breloer 1994)

may be circular on a *pruning wound* or form around the perimeter of a *canker*.

Wound margin entire The *wound wood* growth in the margin is mostly uninterrupted forming a uniform edge.

Wound margin irregular The *wound wood* growth in the margin is mostly interrupted and forms an edge that is not uniform, e.g. where repeated wounding *episodes* such as inflicted by ongoing borer activity damages and continually alters the *wound perimeter* with *callus wood* and *wound wood*. See also *Successional wound*.

Wound margin left The left side of a wound margin when the distal and proximal ends of the wound is known, to determine the *wound apex* and *wound base*, respectively.

Wound margin right The right side of a wound margin when the distal and proximal end of the wound is known, to determine the *wound apex* and wound base, respectively.

Wound margin width Distance from *wound margin* to the site of initial wounding. Where evident the *initial wound margin* may be identified by discoloured bark or bark of a different texture to adjacent undamaged trunk. This may also assist in determining the age of a wound.

Wound paints See *Wound dressing*.

Wound perimeter The edges of a wound comprised of its *apex*, *base* and *margins*.

Wound sealants See *Wound dressing*.

Wound wood Aged *callus wood* around the margins of a wound that becomes differentiated to form *CODIT Wall 4* producing new lignified wood. This wood may grow to surround a wound and may eventually develop to enclose the wound by *occlusion*.

Xerophyte A plant adapted to survive in regularly or predominantly dry soil conditions.

Xylem Woody conducting *tissue* with the primary role of the upward *translocation* of water and solutes and is usually internal on a stem to *phloem* being formed at the inner side of the *vascular cambium* as the *primary xylem* and *secondary xylem*.

Xylem rays See *Vascular ray*.

Young Tree aged less than <20% of life expectancy, *in situ*. See also *Age*, *Mature* and *Over-mature*.

Zone of rapid taper The area in the *root plate* where the diameter of *structural roots* reduces substantially over a short distance from the *trunk*. Considered to be the minimum radial distance to provide structural support and *root plate* stability. See also *Structural root zone (SRZ)*.

Zones Land areas identified for particular forms of development by planning laws, i.e. industrial, residential, rural etc.

REFERENCES AND FURTHER READING

Albers J & Hayes E (1993) *How to Detect, Assess and Correct Hazard Trees in Recreational Areas*. Minnesota Department of Natural Resources.

ASTT (Australasian Society for Trenchless Technology) (2003) Trenchless technologies. <http://www.ASTT.com.au>. Accessed January 2003.

Australian Standard (2002) *AS 1348 Road and Traffic Engineering*. Standards Australia, Sydney.

Australian Standard (2007). *AS 4373 Pruning of Amenity Trees*. Standards Australia, Sydney.

Bailey J (Ed.) (1999) *Penguin Dictionary of Plant Sciences*. Penguin, Melbourne.

Barrell J (1993) Preplanning tree surveys: safe useful life expectancy (SULE) is the natural progression. *Arboricultural Journal* **17**(1), 33–46.

Barrell J (1995) Pre-development tree assessments. In: *Trees and Building Sites: Proceedings of an International Conference Held in the Interest of Developing a Scientific Basis for Managing Trees in Proximity to Buildings*. (Eds GW Watson and D Neely) pp. 132–142. International Society of Arboriculture, Savoy, Illinois.

Barrell J (2002) TreeAZ, version 5.03. <http://www.barrelltreecare.co.uk/treeaz/treeaz.asp>. Accessed 27 June 2005.

Beard JS (1990) *Plant Life in Western Australia*. Kangaroo Press, NSW.

Blackman RA (1984) *Penguin Dictionary of Botany*. Allen Lane Penguin Books, London.

Blank P (2006) *Basic Decay Detection Manual for Trees and Timber Structures*. IML Australia, Brisbane.

Boland DJ, Brooker MIH, Chippendale GM, Hall N, Hyland BPM, Johnston RD, Kleinig DA, McDonald MW & Turner JD (2006) *Forest Trees of Australia*. 5th edn. CSIRO Publishing, Melbourne.

British Standard (1991). *BS5837 Guide for Trees in Relation to Construction*. British Standards Institution, London, UK.

Brock J (1993) *Native Plants of Northern Australia*. Rev. edn. Reed New Holland, Sydney.

Brooker MIH and Kleinig DA (1999) *Field Guide to Eucalypts (Vol. 1) South-eastern Australia*. 2nd edn. Bloomings Books, Melbourne.

Buchanan RA (1989) *Bush Regeneration – Recovering Australian Landscapes.* Macarthur Press, Sydney.

Burrows GE (2002) Epicormic strand structure in *Angophora, Eucalyptus* and *Lophostemon* (Myrtaceae) – implications for fire resistance and recovery. *New Phytologist* **153**, 111–131.

Bureau of Meteorology Australia (2008). <http://www.bom.gov.au/lam/glossary/wpagegl.shtml>. Accessed 17 March 2008.

Caldwell MM & Richards JH (1989) Hydraulic lift: water efflux from upper roots improves effectiveness of water uptake by deep roots. *Oecologia* **79**, 1–5.

Capon B (1990) *Botany for Gardeners – An Introduction and Guide.* Timber Press, Portland, Oregon.

City of Port Phillip, Australia (2006). <http://www.portphillip.vic.gov.au/business_planning_permits.html>. Accessed 16 May 2006.

Clark R (2003) *Specifying Trees – A Guide to Assessment of Quality Trees.* 2nd edn. NATSPEC/Construction Information, Sydney.

Clarke C (1997) *Nepenthes of Borneo.* Natural History Publications (Borneo), Malaysia.

Coder KD (1998) Root growth control: managing perceptions and reality. In: *The Landscape Below Ground II: Proceedings of an International Workshop on Tree Root Development in Urban Soils, International Society of Arboriculture, Champaign, IL, USA.* (Eds D Neely and GW Watson) pp. 51–81.

Coder KD (2004) American mistletoe – kissing under a parasite. *Arborist News* **13**(6), 37–44.

Coder KD (2004) Lightning damage in trees – the spark of death. *Arborist News* **13**(3), 35–44.

Costello LR, Perry EJ, Matheny NP, Henry JM & Geisel PM (2003) *Abiotic Disorders of Landscape Plants – Diagnostic Guide.* University of California Agricultural National Research Publication, California.

Craul PJ (1992) *Urban Soil in Landscape Design.* John Wiley & Sons, New York.

Craul PJ (1999) *Urban Soils – Applications and Practices.* John Wiley & Sons, New Jersey.

Dobson MC & Moffat AJ (1993) *The Potential for Woodland Establishment on Landfill Sites.* Department of the Environment, HMSO, London, UK.

Draper D (1997) Pseudo-street trees: a concept. *The Australian Arbor Age* **2**(3), 6–10.

Elliot WR & Jones DL (1986) *Encyclopaedia of Australian Plants – Suitable for Cultivation (Vol. 4)*. Lothian Books, Melbourne.

Ennos R (2001). *Trees*. The Natural History Museum, London, UK.

Farm Forest Line (2002). <http://www.farmforestline.com.au/pages/6.1_diameter.html>. Accessed 2006.

Forest Service Maryland Department of Natural Resources (2005). http://www.dnr.state.md.us/forests/programs/urban/. Accessed 2005.

Geiger JR (2004) Is all of your rain going down the drain? *Urban Forest Research* **5**(3), 10–11. USDA Forest Service, Center for Urban Forest Research, California. <http://www.fs.fed.us/psw/programs/cufr/products/newsletters/UF4.pdf#xml=http://www.fs.fed.us/cgi-bin/texis/searchallsites/search.allsites/xml.txt?query=Geiger+rain&db=allsites&id=47c5045f0>. Accessed 10 May 2006.

Gilman E (1997) *An Illustrated Guide to Pruning*. Delmar Publishers, New York.

Gilman E & Lilly S (2002) *Best Management Practice – Tree Pruning*. International Society of Arboriculture, Champaign, Illinois.

Gilman E (2003) Branch-to-branch diameter ratio affects strength of attachment. *Journal of Arboriculture* **29**(5), 291–294.

Grant J (1997) *The Nest Box Book*. Gould League of Victoria, Melbourne.

Grgurinovic CA (1997) *Larger Fungi of South Australia*. Botanic Gardens of Adelaide and State Herbarium and the Flora and Fauna of South Australia Handbooks Committee Adelaide, State Government of South Australia.

Hadlington P (1996) *Australian Termites and Other Common Timber Pests*. 2nd edn. New South Wales University Press, Sydney.

Handreck K & Black N (2002) *Growing Media for Ornamental Plants and Turf*. 3rd edn. University of New South Wales Press, Sydney.

Harden GJ & Murray LJ (2000) *Supplement to Flora of New South Wales (Vol. 1)*. University of New South Wales Press, Sydney.

Harris RW, Clark JR & Matheny NP (2004) *Arboriculture, Integrated Management of Landscape Trees, Shrubs and Vines*. 4th edn. Prentice Hall Publications, New Jersey.

Hayes E (2001) *Evaluating Tree Defects – A Field Guide.* 2nd edn. Safetrees, Minnesota, USA.

Helms J (Ed.) (1998) *The Dictionary of Forestry.* Society of American Foresters, Washington, DC.

Hitchmough JD (1994) *Urban Landscape Management.* Inkata Press, Sydney.

Houghton PD & Charman PEB (1986) *Glossary of Terms Used in Soil Conservation.* Soil Conservation Service New South Wales, Australia.

Howard University Astronomy Glossary (2006). <http://www.google.com.au/ search?hl=enandlr=anddefl=enandq=define:RADARandsa=Xandoi= glossary_definitionandct=title>. Accessed 25 April 2006.

Hubbard W, Latt C & Long A (2006) *Forest Terminology for Multiple-Use Management.* School of Forest Resources and Conservation, Florida Cooperative Extension Service, Institute of Food and Agricultural Sciences, University of Florida, and the Florida Forest Stewardship Program. <http:// edis.ifas.ufl.edu/FR063>. Accessed 2 February 2007.

IACA (2005) Sustainable Retention Index Value. Institute of Australian Consulting Arboriculturists. <www.iaca.org.au>. Accessed 3 February 2006.

James K (2003) Dynamic loading of trees. *Journal of Arboriculture* **29**(3), 165–171.

James K (2005) Dynamic wind loads on trees. <http://isaac.org.au/wott/ Dynamic%20Wind%20Loads%20on%20trees.pdf>. Accessed 25 October 2006.

Jones DL (1996) *Palms in Australia.* 3rd edn. Reed Books, Melbourne.

Jones DL & Elliot WR (1986) *Pests, Diseases and Ailments of Australian Plants.* Thomas C. Lothian, Melbourne.

Kalisz PJ, Stringer JW & Wells RJ (1994) Vertical mulching of trees: effects on roots and water status. *Journal of Arboriculture* **20**(3), 141–145.

Keane PJ, Kile GA, Podger FD & Brown BN (Eds.) (2000) *Diseases and Pathogens of Eucalypts.* CSIRO Publishing, Melbourne.

Lawson M & O'Callaghan D (1995) A critical analysis of the role of trees in damage to low rise buildings. *Journal of Arboriculture* **21**(2), 90–97.

Lonsdale D (1999) *Principles of Tree Hazard Assessment and Management.* Department of the Environment, Transport and the Regions, London UK.

Loudon JC (1853) *An Encyclopaedia of Trees and Shrubs: Being the Arboretum Et Fruticetum Britannicum Abridged: Containing the Hardy Trees and Shrubs of Britain, Native and Foreign, Scientifically Described: with Their Propagation, Culture, and Uses in the Arts.* Longman, Brown, Green & Longmans, University of Michigan, USA. <http://books.google. com.au>. Accessed 25 September 2008.

MacLeod RD & Cram WJ (1996) 'Arboriculture research and information note – forces exerted by tree roots.' 134/96/EXT, pp. 1–7. Arboricultural Advisory and Information Services, UK.

Manion PD (1991) *Tree Disease Concepts.* 2nd edn. Prentice Hall, New Jersey.

Matheny NP & Clark JR (1994) *A Photographic Guide to the Evaluation of Hazard Trees in Urban Areas.* 2nd edn. International Society of Arboriculture, Urbana, Illinois.

Matheny NP & Clark JR (1998) *Trees and Development – A Technical Guide to Preservation of Trees During Land Development.* International Society of Arboriculture, Champaign, Illinois.

Mattheck C & Breloer H (1994) *The Body Language of Trees: A handbook for Failure Analysis.* TSO (The Stationery Office), London, UK.

Mattheck C (1997) *Design in Nature – Learning from Trees.* Trans. Dr William Linnard. Research Center Karlsruhe Institute for Material Research II, Karlsruhe, Germany. Springer-Verlag, Berlin.

Mattheck C (1999) *Stupsi Explains the Tree – A Hedgehog Teaches the Body Language of Trees.* 3rd edn. Trans. Dr William Linnard. Research Center Karlsruhe Institute for Material Research II, Karlsruhe, Germany.

Mattheck C (2004) *The Face of Failure – In Nature and Engineering.* Forschungszentrum Karlsruhe, Karlsruhe, Germany.

Mc Maugh J (1985) *What Garden Pest or Disease Is That?* Lansdowne Press, Sydney.

New South Wales Department of Environment and Conservation (2005) *Aboriginal Scarred Trees in New South Wales – A Field Manual.* NSW Department of Environment and Conservation, Hurstville, NSW.

Nicolotti G & Miglietta P (1998) Using high-technology instruments to assess defects in trees. *Journal of Arboriculture* **24**(6), 297–302.

O'Callaghan D & Lawson M (1995) A critical look at the potential for foundation damage caused by tree roots. In: *Trees and Building Sites:*

Proceedings of an International Conference Held in the Interest of Developing a Scientific Basis for Managing Trees in Proximity to Buildings. (Eds GW Watson & D Neely) pp. 99, 104. International Society of Arboriculture, Savoy, Illinois.

Perry TO (1982) The ecology of tree roots and the practical significance thereof. *Journal of Arboriculture* **8**(8), 197–211.

Recher H, Lunney D & Dunn I (1986) *A Natural Legacy Ecology – Ecology in Australia.* 2nd edn. Maxwell Macmillan Publishing Australia, Sydney.

Reid N (1996) *Managing Mistletoe.* North-West Catchment Management Committee, Tamworth, NSW.

Roberts J, Jackson N & Smith M (2006) 'Tree roots in the built environment, research for amenity trees no. 8.' Department for Communities and Local Government. TSO (The Stationery Office), London, UK.

Robinson N (2004) *The Planting Design Handbook.* University of Gloucestershire, UK.

Rost TL (1984) *Botany – A Brief Introduction to Plant Biology.* 2nd edn. Wiley and Sons, New York.

Rudall P (1992) *Anatomy of Flowering Plants – An Introduction to Structure and Development.* 2nd edn. Cambridge University Press, Cambridge, UK.

Russell T & Cutler C (2003) *The World Encyclopedia of Trees.* Anness Publishing, London, UK.

Scott DH (1912) *An Introduction to Structural Botany.* 8th edn, Part 1. Adam and Charles Black, London, UK.

Shepherd CJ & Totterdell CJ (1988) *Mushrooms and Toadstools of Australia.* Inkata Press, Melbourne.

Shigo AL (1979) 'Tree decay an expanded concept.' Information Bulletin Number 419. United States Department of Agriculture Forest Service. <http://www.na.fs.fed.us/spfo/pubs/misc/treedecay/cover.htm>. Accessed 19 March 2007.

Shigo AL (1986) *A New Tree Biology Dictionary.* Shigo and Trees, Durham, New Hampshire, USA.

Shigo AL (1989a) *A New Tree Biology.* 2nd edn. Shigo and Trees, Durham, New Hampshire, USA.

Shigo AL (1989b) *Tree Basics.* Shigo and Trees, Durham, New Hampshire, USA.

Shigo AL (1989c) *Tree Pruning: A Worldwide Photo Guide*. Shigo and Trees, Durham, New Hampshire, USA.

Shigo AL (1991) *Modern Arboriculture: A Systems Approach to the Care of Trees and Their Associates*. Shigo and Trees, Durham, New Hampshire, USA.

Shigo AL (2004) *Tree Pruning Basics*. Shigo and Trees, Durham, New Hampshire, USA.

Smith SE & Read DJ (1997) *Mycorrhizal Symbiosis*. 2nd edn. Academic Press, London, UK.

Southampton City, UK (2006). <http://www.southampton.gov.uk/environment/development-control/planning-terms.asp>. Accessed 16 May 2006.

Spencer R (1995) *Horticultural Flora of South-Eastern Australia, Volume 1, Ferns, Conifers and Their Allies – The Identification of Garden and Cultivated Plants*. University of New South Wales Press, Sydney and Royal Botanic Gardens, Melbourne.

Stansfield WD (2007) Arbor Day confusion. *Arborist News* **16**(1), 68.

Suzuki D & Grady W (2004) *Tree: A Biography*. Allen and Unwin, Sydney.

Stephen P (2003) The Australian Master Tree Grower Program, University of Melbourne, Australia, Institute of Land and Food Resources Maintained by School of Resource Management. <http://www.mtg.unimelb.edu.au/>. Accessed 2004.

The National Arbor Day Foundation, USA (2005). <http://www.arborday.org/arborday/arborDayDatesinternational.cfm>. Accessed 2006.

Thomas P (2000) *Trees: Their Natural History*. Cambridge University Press, Cambridge, UK.

Thompson WJ & Sorvig K (2008) *Sustainable Landscape Construction – A Guide to Green Building Outdoors*. 2nd edn. Island Press, Washington, USA.

TreeRadar Inc. (2007). <http://www.treeradar.com/TreeRadarFAQs.htm#AboutTreeRadar>. Accessed 10 June 2007.

Trehane P, Brickell CD, Baum BR, Hetterscheid WLA, Leslie AC, McNeill J, Spongberg SA & Vrugtman F (Eds) (1995) *International Code of Nomenclature for Cultivated Plants*. Quarterjack Publishing, Wimborne, UK.

USDA (1979) Forest Service Information Bulletin no. 419.

US Department of Energy (2003) 'A consumer's guide to energy efficiency and renewable energy 2003'. <http://www.eere.energy.gov/consumer/renewable_energy/solar/index.cfm/mytopic=50013>. Accessed 10 June 2007.

Wattle Day Association (2005). <http://www.wattleday.asn.au/mission.html>. Accessed 1 September 2006.

Weber K & Mattheck C (2003) *Manual of Wood Decay in Trees*. The Arboricultural Association, Hampshire, UK.

Wikipedia – The Free Encyclopedia (2007). <http://en.wikipedia.org/wiki/Dendrology>. Accessed 20 February 2007.

Wikipedia – The Free Encyclopedia (2008). <http://en.wikipedia.org/wiki/Standard>. Accessed 10 April 2008.

Willoughby City Council, Australia (2008) 'Willoughby Development Control Plan, Schedule 3, Dictionary of Terms.' <http://202.148.138.211/IgnitionSuite/uploads/docs/Schedule%203%20-%20Dictionary%20of%20Terms.pdf>. Accessed 15 April 2008.

World Wide Wattle (2004) 'Wattle Day.' <http://www.worldwidewattle.com/infogallery/symbolic/wattleday.php>. Accessed 8 September 2004.

Young AM & Smith K (2005) *A Field Guide to the Fungi of Australia*. University of New South Wales Press, Sydney.

TOPICS WITHIN INDEX

INDEX DIVIDED INTO TOPICS

2. Age of trees

3. Animals and habitat in trees

6. Buds

19. Pruning

30. Wounds